Antibody Production

Edited by

L. E. Glynn and M. W. Steward
Canadian Red Cross Memorial Hospital,
Taplow, Bucks.
and
London School of Hygiene and Tropical Medicine
London

JOHN WILEY AND SONS

Chichester · New York · Brisbane · Toronto

British Library Cataloguing in Publication Data:

Antibody production.
  1. Antigens and antibodies
  I. Glynn, Leonard Eleazar
  II. Steward, M. W.   III. Immunochemistry
  574.2′92   QR186.5   80-41378

ISBN 0 471 27916 1

Printed and bound in Great Britain
at The Pitman Press, Bath

List of Contributors

T. J. KINDT   *The Rockefeller University, New York, New York 10021, USA*

R. M. E. PARKHOUSE   *National Institute for Medical Research, Mill Hill, London NW7 1AA, UK*

J. A. SOGN   *The Rockefeller University, New York, New York 10021, USA*

D. R. STANWORTH   *Department of Experimental Pathology, The University of Birmingham, Birmingham B15 2TJ, UK*

A. R. WILLIAMSON   *Department of Biochemistry, University of Glasgow, Glasgow G12 8QQ, UK*

Contents

First published as Chapters 3, 4, 5, and 6 in *Immunochemistry: An Advanced Textbook*, edited by L. E. Glynn and M. W. Steward. © 1977 by John Wiley and Sons Ltd.

Preface

Antibodies are a highly complex group of biologically polyfunctional protein molecules which are produced by conventional protein biosynthetic mechanisms. They possess the unique characteristic of constant and variable amino acid sequences in the same polypeptide chain. This is clearly related to the antigen binding property of the N-terminal region of the chains but the genetic control of the synthesis of these molecules and the origin of the vast diversity of antibody specificities produced by animals and man has posed intriguing questions to the immunologist. This book contains chapters which deal with the biochemical and genetic aspects of antibody production and also a discussion of the immunoglobulinopathies, diseases in which normal antibody production is disturbed. The proteins produced in these diseases have provided material on which much of our understanding of immunoglobulin structure is based and has given vital leads to the mechanism of antibody production.

Undergraduates and graduates in science and medical faculties and research workers in other areas of biological science will find this book a valuable guide to the present state of knowledge in this exciting and challenging field of biology.

CHAPTER 3

Biosynthesis of Immuno-globulins

R. M. E. Parkhouse

1 INTRODUCTION

The production of circulating antibodies by warm-blooded vertebrate animals is the consequence of cell interactions under the stimulating influences of an immunogenic challenge. Although these cell interactions are poorly understood, the major types of cells involved are the B and T lymphocytes (Cooper et al., 1966; Miller and Mitchell, 1969), the latter being capable of aiding or suppressing an immune response (Mitchison, 1971; Gershon, 1975; Tada et al., 1975). Macrophages also may play an accessory role in the response, thereby increasing the complexity of the system. What is clear is that the precursors of high-rate antibody-secreting cells are B lymphocytes, while T lymphocytes never develop into cells which secrete immunoglobulin. This article therefore is confined to the expression and production of immunoglobulins by B lymphocytes and their progeny.

2 STRUCTURE AND GENETICS OF IMMUNOGLOBULINS

The structure and genetics of immunoglobulins are both discussed elsewhere in this volume: structure by Turner (Chapter 1) and genetics by both Sogn and Kindt (Chapter 4) and by Williamson (Chapter 5). However, for clarity and completeness, a brief discussion of these aspects of immunoglobulins will be considered here.

mm mol = 2H, 2L, disulphide bridges.

Immunoglobulins are multichain proteins built up from two basic types of polypeptide chain—heavy and light chains. In its simplest arrangement, an immunoglobulin molecule consists of two heavy chains and two light chains assembled through disulphide bridges. Such a structure is commonly called a monomer and is typical of the IgG, IgD, and IgE classes. Higher molecular weight forms may be constructed by forming disulphide-linked polymers from the monomer subunits. When this occurs, as in the case of the IgM and IgA classes, a further polypeptide, the J chain (Inman and Mestecky, 1974; Koshland, 1975) is required. Typically, IgM is secreted as a polymer consisting of five monomer (IgMs) units, whereas IgA may be secreted in a variety of forms—monomeric, dimeric, trimeric, or tetrameric. However, the important point is that whatever the size of the polymer, only one J chain is incorporated (Chapuis and Koshland, 1974). Furthermore, the immunoglobulin found in mucous secretions (sIgA) is a dimeric form which includes one J chain and a further polypeptide chain termed the secretory component (Tomasi and Grey, 1972).

The distinction between different immunoglobulin classes and subclasses resides in their heavy chains. Nomenclature reflects this fact; thus, the heavy chains of IgG are referred to as γ chains, of IgA as α, of IgM as μ, and so on. Subclasses are defined as, for example in the human, γ_1, γ_2, γ_3, and γ_4. Light chains may be classified into two classes, κ and λ, either of which may be found in random association with any class of heavy chain. A given immunoglobulin molecule is always symmetrical consisting entirely of light chains of the same class and heavy chains of the same class (or subclass).

A fundamental feature of immunoglobulin polypeptide chains became apparent as a result of amino acid sequential studies. In both heavy and light chains, the N-terminal region, comprising about 100 residues, is almost always unique and comprises the variable (V) region (V_H for the heavy chain, V_L for the light chain). The antibody combining site results from interaction between V_H and V_L. Thus, antibody diversity reflects the diversity of variable regions. The remaining C-terminal sequence of the heavy and light chains is invariant and distinct for a given classification, and is therefore called the constant (C) region. The constant regions of light chains, for example, will therefore show one or other invariant sequence characteristic of either κ or λ. Similarly, constant regions of human IgG subclasses may be $C_\gamma1$, $C_\gamma2$, $C_\gamma3$, or $C_\gamma4$.

The fact that immunoglobulin polypeptide chains are composed of C and V regions complicates genetic studies. The major problem is that of how can many V regions be associated with the same C-region sequence, and it is basically this which is responsible for the 'two gene—one polypeptide chain' proposal for immunoglobulins. This concept is well established now and there is considerable experimental evidence to support it. Nonetheless, the majority of genetic studies centre on those genes coding for C regions.

Each defined C region of a heavy or light chain is under the control of a classical Mendelian locus. Genes specifying the various C_H genes (C_μ, $C_\gamma1$, $C_\gamma2$, and so on) are linked closely, but C_κ and C_λ genes are unlinked to C_H genes or to each other. In the rabbit, the V_H and C_H genes are linked closely, and there are recent

indications that a similar situation exists in mice. Strong eviden[
integration of the information for V and C regions at the DNA l[
be discussed below. All C_H-region genes are drawn from the s[
region genes. There are, however, two further pools of V-region g[
and the other for C_λ.

Answers to the questions of the size of the V-region pool and the mechanism of its generation are not known. At best, informed—and sometimes prejudiced—guesses can be made, and arguments still rage as to whether all V regions are carried in the germ line or whether they arise from a few germ-line genes by somatic diversification. It is suggested that the reader refers to reviews by Kindt and Sogn (Chapter 4) and by Williamson (Chapter 5) in this volume, with the warning that the issue is not settled definitively.

For many C-region gene loci there are allelic alternatives and these are referred to as allotypes. An animal may be homozygous or heterozygous with respect to a given allotype system. In heterozygous animals, although the serum contains both allotypic variants, analysis at the level of the individual cell reveals a remarkable specialization, namely that a single cell synthesizes only one of the two allelic alternatives. In an individual cell there is, therefore, exclusive expression of one allele at each immunoglobulin locus. The genetic basis of this phenomena of allelic exclusion is not understood. Nonetheless, the consequence is an expression of one unique heavy–light chain pair, and thus one antibody combining site.

The structure and genetics of immunoglobulin molecules has been the subject of a number of recent reviews to which the reader is referred (Lennox and Cohn, 1967; Herzenberg *et al.*, 1968; Fudenberg and Warner, 1970; Milstein and Munro, 1970; Milstein and Pink, 1970; Pink *et al.*, 1971; Gally and Edelman, 1972; Capra and Kehoe, 1975; Hood *et al.*, 1975; Williamson, 1975).

3 THE LYMPHOCYTE

A characteristic feature of B lymphocytes is the presence of easily detectable immunoglobulin on their surfaces (Raff *et al.*, 1970), and it is this material that acts as the receptor for antigen (Walters and Wigzell, 1970; Strayer *et al.*, 1975). Following this interaction, and perhaps with help from T lymphocytes and macrophages, proliferation and differentiation results in the production of memory cells and high-rate immunoglobulin-secreting cells. The mechanism by which such cells are stimulated to divide and differentiate is still unknown. Equally mysterious is the non-specific, or polyclonal, stimulation of B lymphocytes by mitogens.

It is clear from some recent experiments using the fluorescence-activated cell sorter (FACS) that allotype exclusion occurs in normal lymphocytes (Jones *et al.*, 1974a). Lymphocytes from rabbits heterozygous at the b allotype locus were reacted with fluorescent-labelled, monospecific anti-allotype antibodies, and then were separated into fluorescent and nonfluorescent populations. When tested either in an *in vivo* transfer system, or by stimulation with pokeweed antigen *in vitro*, lymphocytes with a particular b locus allotype on their surfaces give rise to cells secreting the same (and no other) b locus allotype. Earlier suggestions that allotypic exclusion may not operate in lymphocytes are due, presumably, to problems associated with passive absorption by lymphocytes of immunoglobulin from serum.

Contentious issues are the number and physiological role of lymphocytes bearing IgG or IgA on their surfaces. Biochemical analysis (Abney and Parkhouse, 1974; Parkhouse *et al.*, 1976; Abney *et al.*, 1976; Melcher *et al.*, 1974; Vitetta *et al.*, 1975) indicates that there is little IgG or IgA expressed on the surface of murine lymphocytes. Instead, the major classes represented are IgD and IgM. The available evidence suggests that a similar situation exists in man and that many cells apparently bearing surface IgG do so as a result of absorption from the serum (Kurnick and Grey, 1975; Winchester *et al.*, 1975). It therefore appears that the number of lymphocytes actually synthesizing membrane IgG and IgA is small, and their exact role in immune responses is not clear.

A crucial question which remains to be resolved is if IgG and IgA-secreting cells are derived from lymphocytes with those immunoglobulins present on their surfaces. At present there is no firm evidence to exclude the derivation of at least some cells secreting IgA and IgG from lymphocytes bearing IgM and/or IgD. In fact, there is evidence that, in the rabbit, there are precursors of IgA-secreting cells which have neither IgA nor IgM on their surfaces (Jones *et al.*, 1974*b*). It seems likely that, in this case, the cell-surface immunoglobulin will prove to be the rabbit equivalent of IgD. Furthermore, the failure to suppress expression of IgG by treatment of lymphocytes with antisera to γ-chain determinants (Mage, 1975) may simply reflect the fact that precursors of IgG-secreting cells do not have IgG on their surface.

An intriguing observation is the simultaneous presence of IgM and IgD on the surface of the same lymphocyte in humans (Knapp *et al.*, 1973; Rowe *et al.*, 1973) and mice (Parkhouse *et al.*, 1976; Abney *et al.*, 1976). It is of particular interest that in such cells the V region is identical in both immunoglobulin classes (Pernis *et al.*, 1974; Salsano *et al.*, 1974; Fu *et al.*, 1975). This observation raises an important question, the solution of which will have implications for the mechanism of V–C-gene integration. The question is: Does the simultaneous expression of two C-region genes—C_μ and C_δ—sharing one V-region sequence result from the presence of two integrated heavy-chain genes (V–C_μ and V–C_δ) in the chromosome? An alternative explanation could be based on the persistence of mRNA for μ chain, for example, in a cell containing only the integrated V–C gene. The former possibility demands the simultaneous integration of C_μ and C_δ genes with the same V-region gene, while in the latter case, there would be successive gene integration events. Should simultaneous integration be demonstrated for C_μ and C_δ, then the possibility of simultaneous integration of all C_H regions is raised. Resolution of this problem requires the demonstration of mRNA synthesis for immunoglobulin heavy chains in those cells with cell-surface IgM and IgD. However, the fact that many long-lived chronic lymphatic leukaemic cells do have IgM and IgD present simultaneously on their surfaces

(Salsano *et al.*, 1974; Fu *et al.*, 1975) does suggest that two integrated heavy-chain genes exist in these cells.

By incubating lymphocytes with radiolabelled amino acids or sugars, it has been possible to demonstrate unequivocally the synthesis of surface immunoglobulin (Melchers and Andersson, 1973; Vitetta and Uhr, 1973, 1974; Parkhouse, 1975). These observations and other studies using cells externally labelled with ^{125}I (Cone *et al.*, 1971; Vitetta and Uhr, 1972; Melchers *et al.*, 1975), also indicate that there is turnover of membrane immunoglobulin. Loss of immunoglobulin *in vitro* occurs by a process termed 'shedding' (Vitetta and Uhr, 1972), since the immunoglobulin released from lymphocytes has been found complexed to fragments of plasma membrane. Whether or not this process occurs *in vivo* is uncertain. The analysis of surface immunoglobulin turnover is complicated by the heterogeneity of lymphocyte populations usually used in these studies. In a recent review (Melchers *et al.*, 1975), have divided lymphocytes into two subpopulations. One cell-type is large and releases cell-surface immunoglobulin with a half life of one to three hours. On the other hand the second subpopulation is small and releases surface immunoglobulin at an appreciably slower rate; the half life is 20–28 hours. It has been suggested that the smaller cell is derived from the larger (Melchers *et al.*, 1975). Clearly, the distribution of IgM and IgD on these different cell populations should be determined.

Unlike the plasma cell, the lymphocyte is not characterized by the secretory apparatus of intracellular membranes so typical of other secretory cells. The question of how surface-membrane constituents, including immunoglobulins, reach their final site in the plasma membrane is not well understood. It is possible (Vitetta and Uhr, 1974), but certainly not proven, that surface immunoglobulin is translated and transported within the cell on membrane elements, as is the case in plasma cells. What is interesting is that immunoglobulin may be secreted, as it is by plasma cells, or may be associated with the surface membrane as in the case of lymphocytes. It is reasonable to ask, therefore, if there is anything to distinguish between membrane-associated and secreted immunoglobulin. There are two obvious differences. One is that a large proportion of surface immunoglobulin is IgD, an immunoglobulin which at best is only a minor component of serum. Secondly, IgM found on the surface of B lymphocytes is in the monomeric form (IgMs), whereas secreted IgM is generally in the pentameric form (Vitetta *et al.*, 1971; Marchalonis *et al.*, 1972).

Unfortunately, these differences do not help to explain the mode of insertion of immunoglobulin into cell membranes. In the absence of sequential data for IgD, it is impossible to say whether its association with cell membranes is correlated with particular structural characteristics, for example a hydrophobic C-terminal sequence. However, the sequence of IgM known (Putnam *et al.*, 1973) and does not indicate C-terminal hydrophobicity. Furthermore, under certain conditions—admittedly pathological in the human (Solomon and Kunkel, 1967) but not so in elasmobranchs—(Marchalonis and Edelman, 1965) monomeric IgMs is secreted into the serum. A suggestion that surface IgM is deficient in galactose and fucose (Melchers and Andersson, 1973) has been contradicted subsequently (Vitetta and Uhr, 1974). At present, small structural differences between surface and secreted immunoglobulin cannot be ruled out, but their demonstration constitutes a formidable task. The actual interaction of surface immunoglobulin with the membrane is another intriguing feature about which little is known. The solution of this problem could well have implications for the mechanism of antigenic stimulation.

4 THE PLASMA CELL

Mature plasma cells are specialized in their cellular organization and the uniqueness of their secretory product. Like most secretory cells, the plasma cell has a highly organized endoplasmic reticulum consisting of both rough and smooth membranes. The rough membrane, so-called because it is studded with ribosomes on the cytoplasmic surface, comprises an extensive network of tubules (cysternae) within the cytoplasm. The smooth, often called the golgi zone, is a mass of vacuoles usually situated towards one pole of the cell. Under normal circumstances, the plasma cell represents a terminal differentiation stage. The immunoglobulin which it secretes is homogeneous and contains the identical V_H–V_L pair expressed on the original precursor lymphocyte. Thus, the antibody specificity of the lymphocyte is preserved in its progeny, exactly as would be predicted from Burnet's theory of clonal selection. A given plasma cell expresses only one C_H and C_L, but examples of plasma cell which secrete one or other of all C-region sequences can be found in the general population (Mäkelä and Cross, 1970). If plasma cells secreting IgG and IgA are derived from lymphocytes with IgM and/or IgD, then there must be a switch in immunoglobulin gene expression during the development of such clones. Experiments in which administration of anti-IgM antibodies *in vivo* abolishes the appearance of IgG and IgA strongly suggest the existence of this sort of phenotypic switch (Kincade *et al.*, 1970).

For many biochemical approaches, the cellular heterogeneity of lymphoid organs poses a major problem. Fortunately, it is possible experimentally to produce myeloma tumours in mouse which can be propagated by serial transplantation (Potter, 1967). These tumours are composed of plasma cells which, unlike their naturally occurring counterparts, divide vigorously. Each myeloma represents the product of a single neoplastic cell, and is therefore a clone of plasma cells secreting a unique homogeneous immunoglobulin. Most studies on the biosynthesis of immunoglobulins have been carried out using these tumour cell lines.

4.1 Synthesis and Assembly of Four-Chain (H_2L_2) Molecules

It is now well established that immunoglobulin heavy and light chains are synthesized independently on polyribosomes in immune tissue and myeloma cells (Scharff and Uhr, 1965; Askonas and Williamson, 1966; Bevan *et al.*, 1972). This is consistent with genetic evidence, which indicates non-linkage between genes for heavy and light chains. The polyribosomes responsible for heavy and light-chain synthesis sediment at about 300 S and 200 S, respectively. Direct visualization of the polyribosomes by electron microscopy (De Petris, 1970) indicates that heavy and light-chain polyribosomes are composed of 11–18 and four to five ribosomes, respectively. As would be expected, experiments in which myeloma cells are exposed to radiolabelled amino acids for short periods demonstrate sequential growth of the polypeptide chain from the N-terminus (Lennox et al., 1967). The actual time of synthesis is short, being about one minute for a light chain and two minutes for a heavy chain.

Given two genes for one polypeptide chain, integration could be at the DNA, RNA, or protein level. The demonstration of sequential growth of heavy and light chains from the N to the G terminus, however, rules out the last possibility (Lennox et al., 1967). A similar conclusion was drawn from experiments with cell-free systems and a 9–13 S fraction of RNA isolated from a light chain-producing myeloma tumour (Stavnezer and Huang, 1971). An RNA of this size would be sufficient to code for an entire light chain. When added to a cell-free system derived from rabbit reticulocytes, the product synthesized gives pepetides almost identical to the light chains synthesized by the myeloma tumour in vivo. The most compelling evidence, however, comes from the work of Milstein and his colleagues (1974a) who were able to isolate the mRNA for a light chain in a sufficiently pure form to establish conclusively that it exists as a single molecule. In addition to the poly A sequence of 200 residues and the bases required for specifying the amino acid sequence of the light chain, there are two untranslated base sequences. One of 150 residues is at the 5'-end preceding the V-region sequence, and the other of 200 residues is at the 3'-end interposed between the C-region sequence and the poly A. The function of these untranslated sequences is not known. One obvious possibility is that part of the untranslated sequence ensures that translation occurs on membrane-bound polyribosomes. It is equally possible, however, that the 'extra' RNA sequences control more generalized functions, such as transport from the nucleus. Finally, somatic cell hybrids formed between xenogeneic (Cotton and Milstein, 1973) or syngeneic (Köhler and Milstein, 1975) immunoglobulin-synthesizing cells only produce the parental heavy or light chains. In other words, there is no evidence for scrambling of the C and V regions of heavy and light chains contributed by each of the parental cell lines. Thus, V–C integration cannot possibly result from cytoplasmic events.

The sum total of evidence therefore, strongly points to integration of V and C genes at the level of DNA. What cannot be ruled out formally at present is integration during the transcription of separate V and C genes. This possibility is rather unlikely, however, since there are plasma cells which secrete abnormal heavy chains with part of both the C and V regions deleted (heavy-chain disease) (see Stanworth, Chapter 6). It is difficult to see how such proteins could arise given an integration event at the RNA level. The most probable explanation for heavy-chain disease proteins is that a deletion occurs during or after fusion of V and C genes. On the other hand, it has to be admitted that a suitably large deletion could produce the same result even if the V and C genes were not integrated.

The major impetus for the isolation of mRNA molecules for immunoglobulin is their use as probes for nucleic acid hybridization studies in order to measure directly the size of the V-gene pool. This aspect is discussed by Williamson in this volume (Chapter 5). In the course of establishing the purity of the mRNA (a prerequisite for all hybridization studies), it is usual to examine the product formed when the mRNA is used to programme translation in a cell-free system. A number of workers have made the same interesting observation (Milstein *et al.*, 1972; Swan *et al.*, 1972; Mach *et al.*, 1973; Tonegawa and Baldi, 1973; Schmeckpeper *et al.*, 1974; Green *et al.*, 1975). They have found that the translation product of light-chain mRNA in a cell-free system is larger than the secreted light chain. This enlarged light chain (pro-light chain) is thought to be a precursor to the secreted form. It contains an extra 15–20 amino acids at the N-terminus which are cleaved presumably *in vivo*. The biological significance of the precursor form is not understood, but one suggestion by Milstein *et al.* (1972) is that the extra N-terminal sequence could determine that translation takes place on membrane-bound ribosomes.

In the case of the heavy chain the product of the cell-free system is, if anything, slightly smaller than the secreted form, but it has been argued that a similar situation exists. It has been suggested that, although there is a precursor segment, heavy chain translated in the cell-free system is devoid of carbohydrate (Cowan and Milstein, 1973; Cowan *et al.*, 1973; Green *et al.*, 1975). These two factors would essentially cancel each other out with the consequent similarity in mobility between precursor and secreted heavy chains.

In many, but not all, cases, the synthesis of heavy and light chains is balanced, so that the only secretory product of plasma cells is fully assembled immunoglobulin (Askonas and Williamson, 1969; Baumal and Scharft, 1973a). It would seem likely that this balance is achieved at the transcriptional rather than at the translational level.

The immediate event following the completion of the synthesis of heavy and light chains is their assembly into four-chain structures. In the case of IgA and IgM synthesis, this is followed by the addition of the J chain and the formation of polymeric structures, while for the class of IgA found in the mucosal secretions (sIgA) another polypeptide, the secretory component is added (Tomasi and Grey, 1972). The addition of secretory component takes place after secretion of dimeric IgA molecule from plasma cells, secretory component being synthesized by epithelial cells rather than by plasma cells. Assembly is accompanied or followed by almost invariably the addition of carbohydrate at various points on the polypeptide chain.

In the majority of species in which biosynthetic assembly has been studied, the four-chain structures are held together by both covalent disulphide bridges and non-covalent linkages. It is experimentally feasible to study the former (but not the latter) and to define the order of disulphide bridge formation. This proves to be variable for immunoglobulins as a whole, but is characteristic and constant for a given immunoglobulin class or subclass (Bevan *et al.*, 1972; Baumal and Scharff, 1973a). The first disulphide-linked intermediate formed in the assembly of an H_2L_2 structure can be either H_2 or H–L. Conversion to the four-chain molecule then takes place either by addition of light chains to H_2 or by dimerization of H–L.

In practice, cell suspensions are incubated for relatively short times and then intracellular immunoglobulin and immunoglobulin subunits are isolated and characterized by gel electrophoresis. Absolutely essential, however, is the demonstration that a given Ig subunit is an intermediate in the biosynthesis of the secreted immunoglobulin. To achieve this, pulse–chase experiments are carried out, and then the flow of label from one intermediate to another can be shown. From this type of analysis, it is clear that covalent assembly to the H_2L_2 structure is, like synthesis of the polypeptide chain, relatively rapid, taking between five and 15 minutes.

In the mouse, most IgG is formed *via* the H_2 intermediate, whereas IgM is assembled from HL subunits. The absence of disulphide bonds between the heavy and light chains of mouse IgA means that H_2 is the only possible disulphide bonded intermediate. Interestingly, the pattern of disulphide bridge formation within the cell correlates with the lability of the disulphide bonds to reducing conditions (Bevan *et al.*, 1972). Thus, in molecules which assemble via H–L intermediates the H–H disulphide bond is more susceptible to cleavage by reduction than is the H–L disulphide bridge and *vice versa*. Thus, by simply studying the reduction intermediates of an immunoglobulin, it is possible to predict the biosynthetic pathway.

One should not be misled into thinking that the order of disulphide bond formation between the chains is necessarily the same as the order of interactions of the polypeptide chains. The order of non-covalent chain interaction is more difficult to define. However, a feature of immunoglobulin heavy and light chains is their strong non-covalent interactions, which are strong enough to give a functional, antigen-binding H_2L_2 structure in the absence of covalent linkages. On this basis, then, given a mixture of heavy and light chains within the cell, a reasonable prediction would be the immediate formation of non-covalently linked H_2L_2 entitites. The heavy and light chains being appositely placed would, of course, be favourably positioned for formation of inter-chain disulphide bonds.

4.2 Immunoglobulin Secretion and the Addition of Carbohydrate

The plasma cell is a typical secretory cell with a well-developed endoplasmic reticulum. Typically, such cells secrete only immunoglobulin and yet immunoglobulin accounts for only 20–40 per cent of the total protein manufactured by these cells. The problem that arises, therefore, is how to segregate the immunoglobulin for export from those proteins synthesized for intracellular use. A solution to this question came originally from work with other secretory cells such as pancreatic (Jamieson and Palade, 1967a, b) and hepatic tissue (Campbell, 1970). From this work a general mechanism for secretory cells and a functional role for the endoplasmic reticulum has been established.

In general, molecules destined for secretion are translated on polyribosomes associated with the endoplasmic reticulum. The polypeptide chain is then extruded as it is synthesized into the cysternae of the endoplasmic reticulum (vectorial release), from where it passes to the exterior milieu *via* the golgi apparatus. With this knowledge gained from studies made on other cells, it was then a relatively simple matter to confirm that similar events occurred in the secretion of immunoglobulin.

From a series of meticulously conducted experiments, conclusive evidence has been presented to show that immunoglobulin synthesis occurs on polysomes associated with the endoplasmic reticulum (Ciopi and Lennox, 1973). The vectorial release of immunoglobulin peptides into the cysternae of the rough endoplasmic reticulum has also been demonstrated (Vassalli *et al.*, 1967; Bevan, 1971*a*). Data from cell fractionation, coupled with carbohydrate analysis (Melchers, 1969; Uhr and Schenkein, 1970; Melchers, 1971*a*, Choi *et al.*, 1971*a*, *b*), and electron microscopic autoradiography of galactose-labelled immunoglobulin has shown the passage of immunoglobulin from the rough endoplasmic reticulum to the golgi zone (Zagury *et al.*, 1970). Immunoglobulin probably is associated with membranes during the whole of its intracellular life.

Studies of the transit time for an immunoglobulin to pass from the site of synthesis to the exterior yield estimates which vary according to the myeloma tumour used. An average half-life value is 90–150 minutes. However, some molecules spend only 30 minutes within the cell, while others remain inside for two hours or more (Scharff *et al.*, 1967; Melchers, 1970; Parkhouse, 1971*a*; Choi *et al.*, 1971*a*, *b*). This appears to be the consequence of a large mixing pool of immunoglobulin within the rough endoplasmic reticulum. As a result, attempts to follow the passage of immunoglobulin from the rough endoplasmic reticulum to the golgi apparatus are inherently difficult, although this type of experiment has been performed. The data of Choi *et al.* (1971*a*, *b*) suggest that an immunoglobulin molecule spends about two-thirds of its intracellular life within the rough endoplasmic reticulum and the remaining third in the golgi apparatus.

A major problem, still unsolved, is how to explain the remarkable selectivity of the secretion mechanism. Some authors have speculated that secreted proteins have 'transport-recognition sequences' (Schubert and Cohn, 1968*b*), or that carbohydrate may label the protein for secretion (Eylar, 1966; Melchers and Knopf, 1967). However, amino acid sequential studies have not identified a transport sequence and many proteins, including immunoglobulin light chains, are devoid of carbohydrate and yet are still secreted (Winterburn and Phelps, 1972). The simplest and most likely explanation is that mRNAs for secretory proteins contain a non-translatable nucleotide sequence with an affinity for membrane-bound ribosomes or ribosomal subunits. In murine myeloma cells, newly synthesized 60 S ribosomal subunits bind directly to membranes and the 40 S ribosome–mRNA complex then binds to the membrane-bound 60 S particle (Baglioni *et al.*, 1971). There are, therefore, certain constraints on any proposed model.

During passage of immunoglobulin molecules through the membranous elements of the cell, carbohydrate residues are added at discrete intracellular sites. Whether the carbohydrate is a prerequisite for secretion is not proven conclusively but, as mentioned above, the existence of secreted light chains lacking in carbohydrate is certainly an argument against an obligatory role for carbohydrate in secretion.

Studies on the incorporation of radiolabelled monosaccharides into immunoglobulin have been carried out using normal rabbit lymphoid cells (Swenson and Kern, 1968; Cohen and Kern, 1969), mouse myeloma cells (Melchers, 1970, 1971b; Schenkein and Uhr, 1970; Uhr and Schenkein, 1970; Choi et al., 1971a; Parkhouse and Melchers, 1971; Cowan and Robinson, 1973; Della Corte and Parkhouse, 1973a), and mitogen-stimulated B lymphocytes (Melchers and Andersson, 1973). The rates of appearance of radioactive immunoglobulin inside and outside the cell, in the presence or absence of an inhibitor of protein synthesis, indicate that glucosamine and mannose are added close to the time of synthesis of the polypeptide chains as well as later. However, galactose is added later during passage through the golgi apparatus, while sialic acid and fucose are added close to the time of secretion. The one exception is IgM, in which case galactose is added very close to the time of secretion.

It has been suggested that the 'bridge sugar' N-acetylglucosamine is attached to growing immunoglobulin chains on the polyribosomes (Moroz and Uhr, 1967; Sherr and Uhr, 1969), as has been claimed for the glycoproteins synthesized by rat liver (Molnar et al., 1965; Robinson, 1969). The evidence for this is not conclusive as no special precuations were taken to overcome the problem of the ribosomes absorbing completed immunoglobulin chains. The discovery of a case of human myeloma with both Bence–Jones protein and IgG (Edmundson et al., 1968) bears on this point. Although the Bence–Jones protein and light chain isolated from the intact IgG had the same primary amino acid sequence, carbohydrate was only present on the Bence–Jones protein. This observation is difficult to reconcile with the addition of the bridge sugar to the nascent chain, as also is the presence of carbohydrate on the $C_H 2$ region of only one of the two heavy chains comprising rabbit IgG (Fanger and Smyth, 1972).

4.3 Polymeric Immunoglobulin

A major immunoglobulin observed on the B lymphocyte membrane is IgMs, perhaps reflecting its origins in ontogeny and phylogeny. It is interesting to consider how such a large molecule is assembled and secreted by plasma cells while also functioning as a receptor for antigen on lymphocytes. Based on the possibility that the normal process of assembly and secretion might be arrested at some stage in lymphocytes with surface IgMs, this section will concentrate on terminal events in the biosynthesis of polymeric immunoglobulins. An interesting problem which arises is why secreted IgM is a uniform pentameric product whereas IgA is secreted as a heterogeneous mixture of monomer and polymers.

The most characteristic feature in the assembly of murine polymeric immunoglobulins is the conversion of 7 S $(H_2 L_2)$ subunits to polymers just before, or simultaneously with, secretion (Parkhouse and Askonas, 1969; Bevan, 1971b;

Parkhouse, 1971a, b, 1975; Bargellesi et al., 1972; Buxbaum and Scharft, 1973; Della Corte and Parkhouse, 1973a; Parkhouse and Della Corte, 1974). This conclusion is based on the observation that the major species of immunoglobulin found in polymer-secreting plasmacytoma and normal lymphoid cells is the 7 S subunit; the amounts of intracellular polymer are low, if detectable at all. Thus, polymerization of the subunits and exit to the exterior must be closely linked-if not simultaneous events. In man, studies with myeloma cells have shown a similar picture for the secretion of IgA (Buxbaum et al., 1974a), but a considerable number of human myeloma tumour cells secreting IgM contain easily detectable quantities of the 19 S polymeric form (Buxbaum et al., 1974).

Isolated intracellular 7 S IgMs does not polymerize spontaneously unless first treated with a reducing agent, suggesting that the cysteine residues responsible for inter-subunit linkage are blocked within the cell (Askonas and Parkhouse, 1971). Removal of the block, perhaps with the participation of an enzyme-mediating disulphide interchange, may form the basis of a control mechanism for the final polymerization step. Other events that take place at the time of polymerization, just before secretion of polymeric immunoglobulins, are the incorporation of the J chain into the molecule (Halpern and Coffman, 1972; Parkhouse, 1972) and the addition of terminal carbohydrate residues (Parkhouse and Melchers, 1971; Cowan and Robinson, 1972; Melchers, 1972; Della Corte and Parkhouse, 1973a; Melchers and Andersson, 1973; Parkhouse, 1975). The polymerization therefore requires the integration of several defined biochemical events. An interruption at this stage of biosynthesis may establish whether IgM molecules become surface associated rather than being secreted.

Given that the majority of lymphocytes bear surface IgM, key questions to ask concern molecular events following antigenic stimulation. In the study of these processes, the small number of responding cells to a given antigen presents a major problem. However, it is now possible to overcome this difficulty using stimulants that can activate a large number of bone marrow-derived lymphocytes into proliferation and differentiation resulting in IgM-secreting cells (Parkhouse et al., 1972; Melchers and Andersson, 1973). It is to be hoped that biochemical investigations of such inductive systems will lead to an understanding of the molecular mechanism of antigenic stimulation.

In order to explain the control of polymerization, one must consider possible roles for addition of carbohydrate and J chain, an enzyme mediating disulphide interchange, and non-covalent association between subunits.

Carbohydrate does not play a critical role in polymerization because intracellular monomeric forms of IgA and IgM, precursors of the secreted polymers, can be polymerized in vitro (Della Corte and Parkhouse, 1973b). Since intracellular monomeric subunits have been shown to lack fucose and to be deficient in galactose (Parkhouse and Melchers, 1971; Cowan and Robinson, 1972; Melchers, 1972; Della Corte and Parkhouse, 1973a; Melchers and Andersson, 1973; Parkhouse, 1975), addition of these terminal carbohydrate residues is unlikely to be necessary for polymerization. Furthermore, in a murine myeloma secreting monomeric and polymeric forms of IgA, all secreted molecular species have an identical carbohydrate composition (Della Corte and Parkhouse, 1973a).

A mandatory role for J chain (reviewed by Koshland, 1975 and by Inman and Mestecky, 1974) in the assembly of murine polymeric immunoglobulins was shown in a series of experiments in which radiolabelled intracellular or secreted immunoglobulin subunits were polymerized in vitro (Della Corte and Parkhouse, 1973b). Both IgA and IgM can be polymerized from subunits previously prepared by reduction of polymeric forms, and from intracellular or secreted monomeric subunits. Monomeric IgM is converted to the pentamer, and monomeric IgA to the dimer. In all cases, polymerization is total, there being no residue of the monomeric form. Polymerization only occurs when both murine J chain and a purified disulphide-interchange enzyme are available. In the absence of the enzyme, and provided that the concentration of immunoglobulin is relatively high, IgM, but not IgA, can be polymerized from subunits prepared by mild reduction (Parkhouse et al., 1970, 1971; Askonas and Parkhouse, 1971). The importance of the disulphide-interchange enzyme is emphasized therefore by the fact that low concentrations of both IgA and IgM subunits can be polymerized in the presence of the enzyme provided that J chain is supplied.

These experiments infer, but do not demonstrate conclusively, a role for the disulphide-interchange enzyme within the intact secreting cell. While it is true that IgM subunits prepared by reduction will polymerize in vitro in the absence of J chain (Kownatski, 1973; Eskeland, 1974), it is important to note that the product is not pentameric IgM, but, a mixture of molecular sizes. It must be concluded, therefore, that the J chain is essential for the accurate assembly of IgM molecules.

The J chain is not simply a catalyst since all J chain released from IgM by reductive cleavage is incorporated back into reassembled polymer (Della Corte and Parkhouse, 1973b). On the basis of these experiments, the J chain therefore appears to be an essential structural requirement for polymeric immunoglobulins. However, the possibility of rare exceptions to this rule cannot be excluded. Indeed, a human 19 S myeloma IgM (Eskeland and Brandtzaeg, 1974) and certain fish IgM molecules (Weinheimer et al., 1971) appear to be devoid of J chain. However, such rare exceptions do not prove that J chain is not a structural requirement for polymerization of normal IgM. Similarly, the presence of heavy-chain disease myeloma proteins which do not have light chains does not provide proof that light chains are an optional requirement for fully functional immunoglobulin structures.

A remarkable degree of specificity in polymerization has been demonstrated (Della Corte and Parkhouse, 1973b). Not only do reduced albumin and IgG fail to interfere with polymerization, but IgA and IgM subunits can be reassembled simultaneously into specific polymeric forms without the formation of hybrid molecules. Thus, subunits of IgM cannot interact with subunits of IgA, in spite of the fact that the same J chain can mediate polymerization of both immunoglobulin classes.

Assuming that the J chain travels from its site of synthesis by the same route as heavy and light chains, that is through the membranous elements of the cell, then an important factor in the control of polymerization may be the demonstrated lack of measurable non-covalent interactions between J chain and intracellular 7 S subunits. Further control may be due to the fact that the cysteine residues responsible for inter-subunit disulphide bridging are blocked within the cell, suggesting an obvious role for the disulphide-interchange enzyme. High levels of this enzyme are found in the golgi apparatus and plasma membrane, whereas the enzyme is present in the rough endoplasmic reticulum but is inactive (Williams *et al.*, 1968; Della Corte and Parkhouse, 1973*b*). Therefore, during the passage of immunoglobulin subunits through the rough endoplasmic reticulum formation of polymers would not be expected to occur. Within the golgi vesicles, however, the enzyme is present, and so the failure to find appreciable quantities of intracellular polymer is explained perhaps by a short transit time of 7 S subunits through the golgi apparatus. This suggestion gains credence from the low levels of galactose found attached to intracellular monomers (Parkhouse and Melchers, 1971; Melchers and Andersson, 1973; Parkhouse, 1975) and from the fact that galactosyl transferase is located primarily within the golgi apparatus. On the other hand, there is no experimental evidence to rule out a segregation of J chain from 7 S subunits until the assembly site is reached. It is interesting to speculate that the presence of 19 S IgM within some, but not all, human IgM-secreting myeloma cells (Buxbaum *et al.*, 1971) might be due to a prolonged sojourn of such molecules within the golgi apparatus. If this were true, appreciable amounts of galactose-labelled IgM would be expected within the cells. Alternatively, the occurrence of 19 S IgM intracellularly could result from a derangement of a normal control mechanism which segregates the J chain from 7 S subunits.

An intriguing question is why secreted IgM is essentially all uniform pentamer, whereas IgA is secreted as a heterogeneous mixture of the monomer and polymers. Since it is clear that the IgA-producing cells contain the biochemical machinery necessary for polymerization and since secreted (or intracellular) IgA monomer can be polymerized *in vitro*, it is possible that the amount of intracellular J chain could be a critical factor. That this is indeed the case has been demonstrated by the fact that in cells secreting IgM, the synthesis of J chain and 7 S IgM is balanced normally so that neither is produced in excess. In cells secreting IgA, however, there is a deficiency in J chain, resulting in secretion of the monomeric form (Parkhouse and Della Corte, 1973).

Given the fact that J chain is limiting in IgA-secreting cells, formation of larger polymers similar to IgM might appear to offer an economical use of J chain. However, in murine plasma cell tumour MOPC 315, for example, only about 15 per cent of the secreted IgA is accounted for by molecules larger than the dimer (Della Corte and Parkhouse, 1973*a*). Two clues help to rationalize this problem. The first is the finding that J chain is located as a disulphide 'clasp' between only two monomeric subunits in pentameric IgM (Chapuis and Koshland, 1974) and polymeric IgA (Hauptman and Tomasi, 1975). The second is the demonstration of non-covalent interactions between subunits of IgM produced by reduction and alkylation (Tomasi, 1973; Parkhouse, 1974). Partially reduced and alkylated IgM contains material which sediments at 19 S under non-dissociating conditions, but which is dissociated by sodium dodecyl sulphate into oligomeric material (probably the dimer) and 7 S subunits. Thus, alkylated 7 S IgM can associate with oligomeric IgM through non-covalent forces to form a molecule which sediments at 19 S. In contrast, monomeric and dimeric forms of IgA produced by MOPC 315 cells sediment independently. The obvious conclusion is, therefore, that in the polymerization of both IgA and IgM the initial step is the formation of a dimer with the inclusion of J chain. For IgM, further polymerization is promoted by non-covalent interactions between the dimer and monomeric subunits. In the IgA system, such interactions are presumably infrequent, thus further polymerization is not favoured. An explanation for the degree of polymerization, which is characteristic and stable for a given IgA myeloma protein, should then be sought in the intracellular levels of J chain and in the presence or absence of non-covalent interaction between reduced subunits of large polymers.

The exact mechanism of polymerization has been discussed in detail by Koshland (1975) and by Inman and Mestecky (1974). Basically, the proposals involve interaction between J chain and monomeric subunits, and a succession of disulphide-exchange reactions.

An interesting finding has been the detection of J chain in murine (Kaji and Parkhouse, 1974, 1975; Mossman and Baumal, 1975) and human (Brandtzaeg, 1974) plasma cells not secreting polymeric immunoglobulin. Comparable amounts of J chain were found in murine myeloma cells secreting IgG and polymeric IgA or IgM, although it is absent in marine L cells. It is possible that J chains are degraded within IgG-producing cells, since they cannot be detected in culture supernatants of cells incubated for five hours *in vitro*. Furthermore, J chains are also found in murine myeloma cells which synthesize but do not secrete monomeric IgA. Thus the expression of J chain is not related to secretion of polymeric immunoglobulin or even to secretion itself.

What is the significance of J chain present in myeloma cells secreting IgG? While other interpretations must be considered (for example, that the cells studied are neoplastic), it is possible to speculate that intracellular J chain in IgG-secreting cells represents a relic from a previous commitment to IgM synthesis, which would allow the same progenitor cell line to switch from IgG to IgA synthesis. Such a developmental succession of immunoglobulin-class expression has been postulated by Cooper et al. (1972). Accordingly, the genes for heavy and J chain would not be expected to be controlled co-ordinately. This appears to be the case, since J chain has been found in myeloma variants with grossly suppressed synthesis of heavy chains (Kaji and Parkhouse, 1974, 1975; Mosman and Baumal, 1975); in these myelomas the major secretory product is light chains. In a murine myeloma tumour with suppression of both heavy and light-chain synthesis, however, J chain is absent. Whether this means that J and light chain synthesis are under co-ordinate control is difficult to say, due to the aneuploid nature of mouse myeloma cells. More work will have to be done before such a conclusion can be drawn. It is certainly intriguing that J chain is synthesized by cells not producing polymer immunoglobulin, but a full explanation of this observation is not immediately apparent. Interestingly, J chain was not detected in two murine leukaemia cell lines, one T cell-like and the other B cell-like. This stresses the different nature of the stimuli responsible for conversion of lymphocytes to immunological function or neoplastic state.

4.4 Defective Synthesis of Immunoglobulin

The argument of whether the V-gene pool is carried in the germ line or is amplified by somatic variation has prompted a search for variant immunoglobulin arising as a result of mutation in cultures of murine myeloma cells. Thus, a study of somatic mutation in this type of system could help possibly in understanding the generation of antibody diversity. It is equally possible, however, that myeloma cells, being fully differentiated, are not the cells of choice and that precursors of antibody-secreting cells should form the basis of the investigations. This is easy to say but difficult to do, and until there are culture lines of precursor cells there is little hope of tackling this problem.

Given tissue-cultured adapted myeloma cells, the first task is to select variants. Two procedures have been described. In one, myeloma cells are seeded at low cell density in a semi-solid, agar medium (Coffino and Scharff, 1971). Each cell grows into a clonal colony and then the secreted product of each clone is revealed by the application of appropriate antibodies. Most clones continue to secrete immunoglobulin and, as a result, an immune precipitate forms between the clonal product and their anti-immunoglobulin. Some, however, do not secrete immunoglobulin and are recognized by the absence of an immune precipitate using either anti-heavy or anti-light-chain reagents. Others are distinguished by secreting only light chains and this is the most common event observed when IgG-secreting myelomas are screened in this way. The spontaneous loss of heavy-chain expression occurs at the rate of 1×10^{-3} per cell per generation, and this is followed by the loss of light-chain expression at the rate of $4 \cdot 5 \times 10^{-4}$ per cell per generation (Baumal et al., 1973). When mutagens are added to the cell cultures, the rate of change increases and occasionally altered heavy chains (for example with a C-terminal deletion) are found (Preud'homme et al., 1973; Birshtein et al., 1974). In such cases, it is probable that a mutation event is responsible for the phenotypic change. Alternative explanations, such as post-transcriptional defects or activation of a previously unexpressed gene cannot be ruled out formally. Not all non-secreting clones have ceased to synthesize immunoglobulin (Cowan et al., 1974). Both heavy and light chains can be found intracellularly although no secretion takes place. In addition, several examples are available of spontaneously occurring or mutagen-induced myeloma cell variants which synthesize heavy chains in the absence of light chains, but which do not secrete them. The heavy chains may be as large as those of the parental cell, but sometimes they are smaller. In one study by Cowan et al. (1974), the absence of light-chain synthesis was attributed to a defective mRNA for the light chain.

An alternative approach for recognizing variants arising in cultures of myeloma cells has been used by Milstein and his colleagues (Cotton *et al.*, 1973). In this method, the clones are selected randomly and the presence or absence of secreted and intracellular immunoglobulin is screened by isoelectric focusing of cells or culture supernatants of $[^{14}C]$ lysine cultures. The major advantage of this procedure is that small variants in both the V and C regions might be recognized by the screening procedure. Again, non-secreting variants and cells secreting short heavy chains have been identified (Secher *et al.*, 1973; Cowan *et al.*, 1974; Milstein *et al.*, 1974*b*). One major difference in this system is the failure to detect variants secreting only light chains, which is particularly difficult to understand since myeloma tumours secreting either excess light chains or light chains only are observed frequently in mice (Potter, 1972; Baumel and Scharff, 1973*b*) and in man (Osserman and Takatsoki, 1963). The non-secreting variants are interesting as they all contain a heavy chain shorter than the parental type. To discover the exact structural lesion would be a formidable task, since only very small quantities are available for characterization. The altered heavy chains which are secreted result from deletions, either internally or at the C terminus.

Perhaps the most disappointing feature of this work is the failure to find any alterations in the V regions. With more variants available for analysis it is possible that some will arise. In addition, and although the numbers are small, the frequency of mutations that are best accounted for by chain-termination events are much higher than those explained by substitutions; this would not be expected from simple random point mutations. Of course, the detection of one or a few amino acid substitutions in an immunoglobulin molecule is much more difficult than the detection of gross size differences due to deletions; thus, the distinction may be more apparent than real.

Defective immunoglobulin synthesis is not infrequent in cases of multiple myeloma in man (see Stanworth, Chapter 6). As mentioned previously, one of the most common defects is production of excess light chains. It is arguable whether this represents the normal situation in murine lymphoid tissue; opinions and results vary on this issue. For example, Askonas and Williamson (1967*a, b*), and Askonas (1974) stress that in normal lymphoid tissue the synthesis of heavy and light chains is balanced so that the only secretory product is fully assembled immunoglobulin. In contrast Baumal and Scharff (1973*b*) argue that, while synthesis of heavy and light chains can be balanced in normal lymphoid tissue, there is almost invariably some excess light chain produced. Analysis of a large number of murine myeloma tumours has shown that most secrete excess light chains. Some, however, are balanced with respect to heavy and light-chain synthesis, whereas others synthesize excess light chain but do not secrete it (Baumal and Scharff, 1973*b*). When production of excess light chain does occur in a myeloma tumour it may result from unbalanced synthesis of heavy and light chains within all the cells of the tumour (Parkhouse, 1971*a, b*; Baumal and Scharff, 1973), or from the presence of variants secreting only light chains (Schubert and Cohn, 1968*a*). In normal lymphoid tissue, the absence of cells containing only light chains (Cebra, 1969) would argue that excess light-chain production, when it occurs, results from unbalanced synthesis of heavy and light chains rather than the presence of variants secreting solely light chains.

In many examples of human myelomatosis, expression of genes for heavy chains is suppressed completely and only light chains are produced. These chains are found characteristically in the urine as Bence–Jones protein. Similarly, myelomas secreting abnormal heavy chains have been described (Seligmann *et al.*, 1969; Frangione and Franklin, 1973).

5 HYBRID CELLS

Hybridization of immunoglobulin-producing cells offers the possibility of analysing some of the problems of gene expression. For example, it may be possible to show unequivocally that V and C genes cannot be transcribed on separate mRNAs (Cotton and Milstein, 1973; Köhler and Milstein, 1975). In general, somatic cell hybrids formed between immunoglobulin-producing cell lines and fibroblasts produce little, if any, immunoglobulin (Periman, 1970; Coffino *et al.*, 1971). At present one cannot conclude that immunoglobulin synthesis has been switched off under the influence of the fibroblast partner in the hybrid. Careful karyotype analysis is required to ensure that the myeloma chromosome(s) carrying the information for immunoglobulin is present in the hybrid cell. When murine myeloma cells have been fused with peripheral human lymphocytes not forming detectable immunoglobulins, the resultant hybrid secrete both murine and human immunoglobulins (Schwaber and Cohen, 1974). Assuming that the human contribution to the hybrids is definitely non-secreting lymphocytes (which has not been established critically), then lymphocytes can be turned on to immunoglobulin synthesis by fusion with high-rate immunoglobulin-secreting cells.

Fusion of immunoglobulin producing cells in xenogeneic (Cotton and Milstein, 1973), allogeneic (Bloom and Nakamura, 1974) or syngeneic (Köhler and Milstein, 1975) combinations give hybrid cells which secrete immunoglo-bulin of both parental types. Analysis of the products shows that, while random assortment of the heavy and light chains can occur, there is no evidence for scrambling V and C regions.

The most spectacular hybridization system is one in which spleen cells from mice primed to sheep erythrocytes are hybridized with murine myeloma cells from the same strain (BALB/c) (Köhler and Milstein, 1975). Hybrids are formed at quite high frequency and secrete, in addition to the myeloma protein, specific antibody to the sheep red blood cells. Although most, if not all, heavy-chain classes are represented in the antibody secreted by a mixture of hybrids, upon cloning only one isotope is expressed by a given isolate. There is, therefore, the possibility of preparing hybrid clones secreting specific antibody of all the known heavy-chain classes. Furthermore, since the fusion has been conducted between syngeneic cells, one of which is a myeloma cell, the hybrid can be transplanted serially in mice. Köhler and Milstein (1975) noted that it should be a relatively simple matter to prepare similar hybrids which do not secrete the myeloma protein, either by using a non-immunoglobulin-producing myeloma variant for hybridization or by selecting hybrids which do not express the myeloma protein. In addition, the antibody specificity can, of course, be selected simply by what is injected into the mouse prior to preparation of the spleen cells for the hybridization.

REFERENCES

Abney, E. R., Hunter, I. R., and Parkhouse, R. M. E. (1976). *Nature (Lond.)*, in press.
Abney, E. R., and Parkhouse, R. M. E. (1974). *Nature (Lond.)*, **252**, 600.
Askonas, B. A. (1974). *Ann. Immunol. (Inst. Pasteur)*, **125**, 253.
Askonas, B. A., and Parkhouse, R. M. E. (1971). *Biochem. J.*, **123**, 629.
Askonas, B. A., and Williamson, A. R. (1966). *Proc. Roy. Soc., Ser. B*, **166**, 232.
Askonas, B. A., and Williamson, A. R. (1967a). *Nobel Symposium*, Vol. **3**, (Killander, J., ed.), John Wiley and Sons, London, p. 369.
Askonas, B. A., and Williamson, A. R. (1967b). *Cold Spring Harbor Symp. Quant. Biol.*, **32**, 223.
Askonas, B. A., and Williamson, A. R. (1969). *Antibiot. Chemother.*, **15**, 64.

Baglioni, C., Bleiberg, I., and Zauderer, M. (1971). *Nature, New Biol.*, **232**, 8.
Bailey, L. K., Hannestad, K., and Eisen, H. N. (1973). *Fed. Proc.*, **32**, 1013.
Bargellesi, A., Periman, P., and Scharff, M. D. (1972). *J. Immunol.*, **108**, 126.
Baumal, R., Birshtein, B. K., Coffino, P., and Scharff, M. D. (1973). *Science, N.Y.*, **182**, 164.
Baumal, R., and Scharff, M. D. (1973a). *Transplant. Rev.*, **14**, 163.
Baumal, R., and Scharff, M. D. (1973b). *J. Immunol.*, **111**, 448.
Bevan, M. J. (1971a). *Biochem. J.*, **122**, 5.
Bevan, M. J. (1971b). *Eur. J. Immunol.*, **1**, 133.
Bevan, M. J., Parkhouse, R. M. E., Williamson, A. R., and Askonas, B. A. (1972). *Prog. Biophys. Mol. Biol.*, **25**, 131.
Birshtein, B. K., Preud'homme, J.-L., and Scharff, M. D. (1974). *Proc. Nat. Acad. Sci., U.S.A.*, **71**, 3478.
Bloom, A. D., and Nakamura, F. T. (1974). *Proc. Nat. Acad. Sci., U.S.A.*, **71**, 2689.
Brandtzaeg, P. (1974). *Nature, (Lond.)*, **252**, 418.
Buxbaum, J. N., and Scharff, M. D. (1973). *J. Exp. Med.*, **138**, 278.
Buxbaum, J. N., Zolla, S., Scharff, M. D., and Franklin, E. C. (1971). *J. Exp. Med.*, **133**, 1118.
Buxbaum, J. N., Zolla, S., Scharff, M. D., and Franklin, E. C. (1974). *Eur. J. Immunol.*, **5**, 367.

Campbell, P. N. (1970). *FEBS Letters*, **7**, 1.
Capra, J. D., and Kethoe, J. M. (1975). *Adv. Immunol.*, **20**, 1.
Cebra, J. J. (1969). *Bacteriol. Rev.*, **33**, 159.
Chapuis, R. M., and Koshland, M. E. (1974). *Proc. Nat. Acad. Sci., U.S.A.*, **71**, 657.
Choi, Y. S., Knopf, P. M., and Lennox, E. S. (1971a). *Biochemistry*, **10**, 659.
Choi, Y. S., Knopf, P. M., and Lennox, E. S. (1971b). *Biochemistry*, **10**, 668.
Ciopi, D., and Lennox, E. S. (1973). *Biochemistry*, **12**, 3211.
Coffino, P., and Scharff, M. D. (1971). *Proc. Nat. Acad. Sci., U.S.A.*, **68**, 219.
Coffino, P., Knowles, B., Nathenson, S. G., and Scharff, M. D. (1971). *Nature, New Biol.*, **231**, 87.
Cohen, H. J., and Kern, M. (1969). *Biochim. Biophys. Acta*, **188**, 255.
Cone, R. E., Marchalonis, J. J., and Rolley, R. T. (1971). *J. Exp. Med.*, **134**, 1373.

Cooper, M. D., Lawton, A. R., and Kincade, P. W. (1972). *Clin. Exp. Immunol.*, **11**, 143.
Cooper, M. D., Peterson, R. D., South, M. A., and Good, R. A. (1966). *J. Exp. Med.*, **123**, 75.
Cotton, R. G. H., and Milstein, C. (1973). *Nature*, *(Lond.)*, **244**, 42.
Cotton, R. G. H., Secher, D. S., and Milstein, C. (1973). *Eur. J. Immunol.*, **3**, 135.
Cowan, N. J., Harrison, T. M., Browlee, G. G., and Milstein, C. (1973). *Biochem. Soc. Trans.*, **1**, 1247.
Cowan, N. J., and Milstein, C. (1973). *Eur. J. Biochem.*, **36**, 1.
Cowan, N. J., and Robinson, G. B. (1972). *Biochem. J.*, **126**, 751.
Cowan, N. J., Secher, D. S., and Milstein, C. (1974). *J. Mol. Biol.*, **90**, 691.

Della Corte, E., and Parkhouse, R. M. E. (1973a). *Biochem. J.*, **136**, 589.
Della Corte, E., and Parkhouse, R. M. E., (1973b). *Biochem. J.*, **136**, 597.
De Petris, S. (1970). *Biochem. J.*, **118**, 385.

Edmundson, A. B., Sheber, F. A., Ely, K. R., Simonds, N. B., Hutson, N. K., and Rossiter, J. L. (1968). *Archs. Biochem. Biophys.*, **127**, 725.
Eskeland, T. (1974). *Scand. J. Immunol.*, **3**, 757.
Eskeland, T., and Brandtzaeg, P. (1974). *Immunochemistry*, **11**, 161.
Eylar, E. H. (1966). *J. Theoret. Biol.*, **10**, 89.

Fanger, M. W., and Smyth, D. G. (1972). *Biochem. J.*, **127**, 767.
Frangione, B., and Franklin, E. C. (1973). *Semin. Haematol.*, **10**, 53.
Fu, S. M., Winchester, R. J., and Kunkel, H. G. (1975). *J. Immunol.*, **114**, 250.
Fudenberg, H. H., and Warner, N. L. (1970). *Adv. Human Genet.*, **1**, 131.

Gally, J. A., and Edelman, G. M. (1972). *Ann. Rev. Genet.*, **6**, 1.
Gershon, R. K. (1975). *Transplant. Rev.*, **26**, 170.
Green, M., Graves, P. N., Zehavi-Willner, T., McInnes, J., and Pestka, S. (1975). *Proc. Nat. Acad. Sci., U.S.A.*, **72**, 224.

Halpern, M. S., and Coffman, R. L. (1972). *J. Immunol.*, **109**, 674.
Hauptman, S. P., and Tomasi, T. B. (1975). *J. Biol. Chem.*, **250**, 3891.
Herzenberg, L. A., McDevitt, H. O., and Herzenberg, L. A. (1968). *Ann. Rev. Genet.*, **2**, 209.
Hood, L., Campbell, J. M., and Elgin, S. C. R. (1975). *Ann. Rev. Genet.*, **9**, 305.

Inman, F. P., and Mestecky, J. (1974). In *Contemporary Topics in Immunochemistry*, Vol. 3, (Ada, G. L., ed.), Plenum Press, New York, p. 111.

Jamieson, J. D., and Palade, G. E. (1967a). *J. Cell. Biol.*, **34**, 577.
Jamieson, J. D., and Palade, G. E. (1967b). *J. Cell. Biol.*, **34**, 597.
Jones, P. P., Cebra, J. J., and Herzenberg, L. A. (1974a). *J. Exp. Med.*, **139**, 581.
Jones, P. P., Craig, S. W., Cebra, J. J., and Herzenberg, L. A. (1974b). *J. Exp. Med.*, **140**, 452.

Kaji, H., and Parkhouse, R. M. E. (1974). *Nature*, *(Lond.)*, **249**, 45.
Kaji, H., and Parkhouse, R. M. E. (1975). *J. Immunol.*, **114**, 1218.
Kinkade, P. W., Lawton, A. R., Buckman, D. E., and Cooper, M. D. (1970). *Proc. Nat. Acad. Sci., U.S.A.*, **67**, 1918.
Knapp, W., Bolhuis, R. L. H., Radl, J., and Hijmans, W. (1973). *J. Immunol.*, **111**, 1295.
Köhler, G., and Milstein, C. (1975). *Nature*, *(Lond.)*, **256**, 495.
Koshland, M. E. (1975). *Adv. Immunol.*, **20**, 41.
Kownatski, E. (1973). *Immunol. Commun.*, **2**, 105.

Kurnick, J. T., and Grey, H. M. (1975). *J. Immunol.*, **115**, 305.

Lennox, E. S., and Cohn, M. (1967). *Ann. Rev. Biochem.*, **36**, 365.
Lennox, E. S., Knopf, P. M., Munro, A. J., and Parkhouse, R. M. E. (1967). *Cold Spring Harbor Symp. Quant. Biol.*, **32**, 249.

Mach, B., Faust, C., and Vassalli, P. (1973). *Proc. Nat. Acad. Sci., U.S.A.*, **70**, 451.
Mage, R. G. (1975). *Fed. Proc.*, **34**, 40.
Mäkelä, O., and Cross, A. M. (1970). *Prog. Allergy*, **14**, 145.
Marchalonis, J. J., Cone, R. E., and Atwell, J. L. (1972). *J. Exp. Med.*, **135**, 956.
Marchalonis, J., and Edelman, G. M. (1965). *J. Exp. Med.*, **122**, 601.
Melcher, U., Vitetta, E. S., McWilliams, M., Lamm, M. E., Philips-Quagliata, J. M., and Uhr, J. W. (1974). *J. Exp. Med.*, **146**, 1427.
Melchers, F. (1969). *Biochemistry*, **8**, 938.
Melchers, F. (1970). *Biochem. J.*, **119**, 765.
Melchers, F. (1971*a*). *Biochemistry*, **10**, 653.
Melchers, F. (1971*b*). *Biochem. J.*, **125**, 241.
Melchers, F. (1972). *Biochemistry*, **11**, 2204.
Melchers, F., and Andersson, J. (1973). *Transplant. Rev.*, **14**, 76.
Melchers, F., and Knopf, P. M. (1967). *Cold Spring Harbor Symp. Quant. Biol.*, **32**, 255.
Melchers, F., von Boehmer, H., and Philips, R. A. (1975). *Transplant. Rev.*, **25**, 26.
Miller, J. F. A., and Mitchell, G. F. (1969). *Transplant. Rev.*, **1**, 3.
Milstein, C., and Munro, A. J. (1970). *Ann. Rev. Microbiol.*, **24**, 335.
Milstein, C., and Pink, J. R. L. (1970). *Prog. Biophys. Mol. Biol.*, **21**, 209.
Milstein, C., Adetugbo, K., Cowan, N. J., and Secher, D. S. (1974*a*). In *Progress in Immunology II*, Vol. I, (Brent, L., and Holborrow, J., eds.), North Holland Publishing Co., Amsterdam, p. 157.
Milstein, C., Brownlee, C. G., Cartwright, E. M., Jarvis, J. M., and Proudfoot, N. J. (1974*b*). *Nature, (Lond.)*, **252**, 354.
Milstein, C., Brownlee, G. G., Harrison, T. M., and Mathews, M. B. (1972). *Nature, New Biol.*, **239**, 117.
Mitchison, N. A. (1971). *Eur. J. Immunol.*, **1**, 10.
Molnar, J., Robinson, G. B., and Winzler, R. J. (1965). *J. Biol. Chem.*, **240**, 1882.
Moroz, C., and Uhr, J. W. (1967). *Cold Spring Harbor Symp. Quant. Biol.*, **32**, 263.
Morrison, S. L., and Scharff, M. D. (1975). *J. Immunol.*, **114**, 655.
Mosmann, T., and Baumal, R. (1975). *J. Immunol.*, **115**, 955.

Osserman, E. F., and Takatsoki, K. (1963). *Medicine*, **42**, 357.

Parkhouse, R. M. E. (1971*a*). *Biochem. J.*, **123**, 635.
Parkhouse, R. M. E. (1971*b*). *FEBS Letters*, **16**, 71.
Parkhouse, R. M. E. (1972). *Nature, New Biol.*, **236**, 9.
Parkhouse, R. M. E. (1974). *Immunology*, **27**, 1063.
Parkhouse, R. M. E. (1975). *Transplant. Rev.*, **14**, 131.
Parkhouse, R. M. E., and Askonas, B. A. (1969). *Biochem. J.*, **115**, 163.
Parkhouse, R. M. E.. Askonas, B. A., and Dourmashkin, R. R. (1970). *Immunology*, **18**, 575.
Parkhouse, R. M. E., and Della Corte, E. (1973). *Biochem. J.*, **136**, 607.
Parkhouse, R. M. E., and Della Corte, E. (1974). In *The Immunoglobulin A System*, (Mestecky, J., and Lawton, A. R., eds.), Plenum Press, New York, p. 139.
Parkhouse, R. M. E., Hunter, I. R., and Abney, E. R. (1976). *Immunology*, **30**, 409.
Parkhouse, R. M. E., Janossy, G., and Greaves, M. F. (1972). *Nature, New Biol.*, **235**, 21.
Parkhouse, R. M. E., and Melchers, F. (1971). *Biochem. J.*, **125**, 235.

110

Parkhouse, R. M. E., Virella, G., and Dourmashkin, R. R. (1971). *Clin. Exp. Immunol.*, **8**, 581.
Periman, P. (1970). *Nature, (Lond.)*, **228**, 1086.
Pernis, B., Brouet, J. C., and Seligman, M. (1974). *Eur. J. Immunol.*, **4**, 776.
Pink, J. R. L., Wang, A.-C., and Fudenberg, H. H. (1971). *Ann. Rev. Med.*, **22**, 145.
Potter, M. (1967). *Meth. Cancer Res.*, **2**, 105.
Potter, M. (1972). *Physiol. Rev.*, **52**, 631.
Preud'homme, J.-L., Buxbaum, J. N., and Scharff, M. D. (1973). *Nature, (Lond.)*, **245**, 320.
Putnam, F. W., Florent, G., Paul, C., Shinoda, U., and Shimizu, A. (1973). *Science, N.Y.*, **182**, 287.

Raff, M. C., Sternberg, M., and Taylor, R. B. (1970). *Nature, (Lond.)*, **225**, 553.
Robinson, G. B. (1969). *Biochem. J.*, **115**, 1077.
Rowe, D. S., Hug, K., Forni, L., and Pernis, B. (1973). *J. Exp. Med.*, **138**, 965.

Salsano, P., Froland, S. S., Natvig, J. B., and Michaelsen, T. E. (1974). *Scand. J. Immunol.*, **3**, 841.
Secher, D. S., Cotton, R. G. H., and Milstein, C. (1973). *FEBS Letters*, **37**, 311.
Seligman, M., Mihaesco, E., Hurez, D., Mihaesco, C., Preud'homme, J.-L., and Rambaud, J.-C. (1969). *J. Clin. Invest.*, **48**, 2374.
Scharff, M. D., Shapiro, A. L., and Ginsberg, B. (1967). *Cold Spring Harbor Symp. Quant. Biol.*, **32**, 235.
Scharff, M. D., and Uhr, J. W. (1965). *Science, N.Y.*, **148**, 646.
Schenkein, J., and Uhr, J. W. (1970). *J. Cell. Biol.*, **46**, 42.
Schmeckpeper, B. J., Cory, J., and Adams, J. M. (1974). *Mol. Biol. Reps.*, **1**, 355.
Schubert, D., and Cohn, M. (1968a). *J. Mol. Biol.*, **38**, 263.
Schubert, D., and Cohn, M. (1968b). *J. Mol. Biol.*, **38**, 273.
Schwaber, J., and Cohen, E. P. (1974). *Proc. Nat. Acad. Sci., U.S.A.*, **71**, 2203.
Sherr, C. J., and Uhr, J. W. (1969). *Proc. Nat. Acad. Sci., U.S.A.*, **64**, 381.
Solomon, A., and Kunkel, H. G. (1967). *A. J. Med.*, **42**, 958.
Stavnezer, J., and Huang, R. C. C. (1971). *Nature, New Biol.*, **230**, 172.
Strayer, D. S., Vitetta, E. S., and Köhler, H. (1975). *J. Immunol.*, **114**, 722.
Swan, D., Aviv, H., and Leder, P. (1972). *Proc. Nat. Acad. Sci., U.S.A.*, **69**, 1967.
Swenson, R. M., and Kern, M. (1968). *Proc. Nat. Acad. Sci., U.S.A.*, **59**, 546.

Tada, T., Taniguchi, M., and Takemori, T. (1975). *Transplant. Rev.*, **26**, 106.
Tomasi, T. B. (1973). *Proc. Nat. Acad. Sci., U.S.A.*, **70**, 3410.
Tomasi, T. B., and Grey, H. M. (1972). *Prog. Allergy*, **16**, 81.
Tonegawa, S., and Baldi, I. (1973). *Biochem. Biophys. Res. Commun.*, **51**, 81.

Uhr, J. W., and Schenkein, I. (1970). *Proc. Nat. Acad. Sci., U.S.A.*, **66**, 952.

Vassalli, P., Lisowska-Bernstein, B., Lamm, M. E., and Benacerraf, B. (1967). *Proc. Nat. Acad. Sci., U.S.A.*, **58**, 2422.
Vitetta, E. S., Baur, S., and Uhr, J. W. (1971). *J. Exp. Med.*, **134**, 242.
Vitetta, E. S., Melcher, U., McWilliams, M., Lamm, M. E., Philips-Quagliata, J. M., and Uhr, J. W. (1975). *J. Exp. Med.*, **147**, 206.
Vitetta, E. S., and Uhr, J. W. (1972). *J. Exp. Med.*, **136**, 676.
Vitetta, E. S., and Uhr, J. W. (1973). *Transplant. Rev.*, **14**, 50.
Vitetta, E. S., and Uhr, J. W. (1974). *J. Exp. Med.*, **139**, 1599.

Walters, C. S., and Wigzell, H. (1970). *J. Exp. Med.*, **132**, 1233.

Weinheimer, P. F., Mestecky, J., and Acton, R. T. (1971). *J. Immunol.*, **107**, 1211.

Williams, D. J., Gurari, D., and Rabin, B. R. (1968). *FEBS Letters*, **2**, 133.

Williamson, A. R. (1976). *Ann. Rev. Biochem.*, **45**, 467.

Winchester, R. J., Fu, S. M., Hoffman, T., and Kunkel, H. G. (1975). *J. Immunol.*, **114**, 1210.

Winterburn, P. J., and Phelps, C. F. (1972). *Nature*, (*Lond.*), **236**, 147.

Zagury, D., Uhr, J. W., Jamieson, J. D., and Palade, G. E. (1970). *J. Cell. Biol.*, **46**, 52.

CHAPTER 4

Genetics of Immunoglobulins

J. A. Sogn and T. J. Kindt

1 INTRODUCTION

An individual is capable of synthesizing a large number of antibody molecules, each with a different specificity for one of the many antigens encountered in its lifetime. A major problem in immunogenetics concerns the nature of the genes which direct the synthesis of this large array of molecules. Basic questions concerning the genetic origin of antibody diversity are discussed in this volume by Williamson (see Chapter 5). Although many questions remain unanswered, there exists a body of experimental data which gives an indication of the nature of the

113

antibody genes. In this chapter, studies on the immunogenetic markers of immunoglobulins, allotypes and more recently idiotypes, which have contributed to the majority of these data will be discussed.

Allotypes have been defined as intraspecies antigenic determinants present on immunoglobulins. Many of these serologically detected markers for immunoglobulins have been shown to correlate with variations in amino acid sequence and, therefore, may be considered as markers for primary gene products. Idiotypes, which may be considered antigenic markers for antibody binding sites, have been found in recent studies similarly to correlate with primary structure. With an increase in the amount of sequence data available, non-antigenic, structurally defined markers are being recognized, thus adding significantly to the allotypes and idiotype data.

Studies on allotype inheritance have supplied specific information such as the autosomal codominant nature of immunoglobulin genes (Oudin, 1960a, b) and the absence of linkage relationships between antibody heavy and light-chain allotypes (Dubiski et al., 1962). In addition to this specific information, these studies have yielded results that have implications beyond the field of immunoglobulin inheritance. The presence of common V-region genetic markers on immunoglobulin heavy chains of different classes has opened the possibility that two or more genes interact prior to the synthesis of a complete heavy chain (Feinstein, 1963; Todd, 1963; Kindt and Todd, 1969). These associations of rabbit V_H allotypes with different immunoglobulin classes suggest an exception to the 'one gene—one polypeptide chain' axiom. Similarly, the observation that antibody-producing cells synthesize only one of the two possible immunoglobulin alleles (Pernis et al., 1965; Weiler, 1965) provided the first example of allelic exclusion for an autosomal trait. While allelic exclusion has been observed for X-linked characteristics (Lyon, 1961), these observations suggested a more general occurrence of this phenomenon.

The majority of the immunogenetic studies to be discussed here has been carried out using rabbit and mouse allotypes and idiotypes; these systems will be briefly described. Structural data available concerning allotypic correlates will be emphasized. Detailed reviews covering allotypy in these species have appeared recently (Mage et al., 1973; Kindt, 1975). Much important immunogenetic information has come from studies of human allotypes, but these will not be discussed directly. For a thorough review of human allotypy, the reader is referred to an article by Natvig and Kunkel (1973). Rat allotypes will not be considered, although recent descriptions of the classes, structures, and allotypes of rat immunoglobulins indicate that this system might provide valuable material for future immunogenetic studies (Bazin et al., 1973, 1974a, b; Nezlin et al., 1974).

Linkage relationships between the allotypes and idiotypes within the rabbit or the mouse system will be delineated, and an attempt will be made to integrate these data into a current picture of the nature and interrelationship of the genes controlling antibody synthesis. The phenomenon of allelic exclusion as it exists in immunoglobulin-producing cells will be described and compared to the more fully characterized phenomenon of X-chromosome inactivation in females. A model, proposed by Riggs (1975) for X-chromosome inactivation, will be considered in this context.

The discussion of allotypes as genetic markers normally proceeds under the assumption that the immunoglobulin allotype genes are allelic structural genes. Recently, however, several reports have described deviations from simple allelic behaviour for allotypes (Bosma and Bosma, 1974; Strosberg *et al.*, 1974; Mudgett *et al.*, 1975). Possible explanations for this behaviour will be considered, along with their implications for current concepts of immunoglobulin genetics.

2 EXPERIMENTAL ALLOTYPIC SYSTEMS

Well-established and long-studied allotypic systems exist in man, rabbit, and mouse. Each system has unique features of interest and has provided key information in the development of current concepts of immunogenetics. The ready availability of human Bence–Jones and myeloma proteins at an early stage in the study of immunoglobulin structure and genetics made them the source of many key initial observations. Particularly significant in this respect were the discoveries that heavy and light chains are under independent genetic control (Harboe *et al.*, 1962*a*; Franklin *et al.*, 1962) and that light chains themselves are under the control of at least two genes (Hilschmann and Craig, 1965). Although the human system remains a productive one, technical advances in the ability to manipulate responses in the rabbit and the mouse, combined with the obvious advantages of working with experimental animals, have made the latter systems currently very attractive.

2.1 Allotypes of the Rabbit

At present, the major importance of the rabbit in immunogenetic studies rests on the variety and multiplicity of allotypes which have been described for rabbit immunoglobulins (Mage *et al.*, 1973; Kindt, 1975). Allotypes are available for each region of the molecule with the exception of the V_L region, and preliminary studies indicate that there may be suitable markers for this region (Thunberg *et al.*, 1973). An area of active interest which will be emphasized here is the correlation of allotype with primary structure.

Although rabbit myeloma proteins have not been available for use in structural studies, an adequate substitute became available when injection with bacterial vaccines was shown to induce homogeneous antibody in large quantities (Krause, 1970). Bacterial injections are now being used to obtain homogeneous rabbit antibodies and amino acid sequential data are rapidly becoming available (Chen *et al.*, 1974; Fleischman, 1971; Jaton, 1974*a*, *b*; Margolies *et al.*, 1974). Inbred strains of rabbits have been developed and work has begun on histocompatibility antigens (Chai, 1974; Cohen and Tissot, 1974; Tissot and Cohen, 1974). These developments will enhance the value of the rabbit in future immunogenetic studies.

2.1.1 Immunoglobulins of the Rabbit

Five heavy-chain classes have been described for the rabbit. These are IgG, IgA1, IgA2, IgM, and IgE. There is some evidence for the existence of subclasses of IgG, but these have not been described in detail (Florent *et al.*, 1973).

The rabbit IgG molecule is different from that of human and mouse in several respects. Figure 1 depicts a molecule of rabbit IgG with κ light chains, indicating the disulphide bridges and the various regions of the heavy and light chains. Rabbit light chains may be either κ or λ type; the former comprises more than 90 per cent of the chains in normal circulation. The majority of rabbit κ light chains has seven half-cystine residues (K_B subtype) (Reisfeld *et al.*, 1968). There is a second smaller population of κ chains which has five half-cystine residues (K_A subtype) (deVries *et al.*, 1969; Rejnek *et al.*, 1969; Zikan *et al.*, 1967).

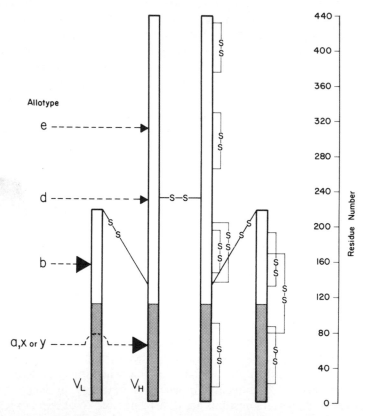

Figure 1 Diagrammatic representation of a rabbit IgG with a K_B light chain. The rabbit light chain differs from the human in having a third disulphide bridge, linking the C_L and V_L domains (Poulsen *et al.*, 1972). The C_HI domain also has an additional intra-chain disulphide bridge and there is only one inter-heavy chain disulphide bond (O'Donnell *et al.*, 1970). The general locations of the known correlates of the various IgG allotypes are shown to the left

Variable region subgroups of κ chains have been described by N-terminal sequential analysis of the chains from homogeneous antibodies to streptococcal carbohydrate (Hood *et al.*, 1970). There is evidence for the existence of at least six κ subgroups (Kindt *et al.*, 1974; Thunberg and Kindt, unpublished results).

2.1.2 Rabbit Allotypes

Rabbit allotypes were first described by Oudin (1956), who coined the name for these intraspecies antigenic determinants. Table 1 summarizes the major allotypic specificities of the rabbit. These specificities will be discussed in more detail below.

Table 1 Major allotypes of rabbit immunoglobulins*

Molecular location	Allotypic group	Allotypes	Correlates
H Chain			
V Region	a	1, 2, 3	complex, see text
	x	32	not known
	y	33	not known
C Region-IgG	d	11, 12	d11-methionine (225)
			d12-threonine (225)
	e	14, 15	e14-threonine (309)
			e15-alanine (309)
C Region-IgA	f	68, 70, 71, 72, 73	not known
C Region-IgA	g	74, 75, 76, 77	not known
C Region-IgM	n	81, 82	not known
κL Chain			
C Region	b	4, 5, 6, 9	complex, see text
λL Chain			
C Region	c	7, 21	not known
Secretory component	t	61, 62	not known

* For complete references, see Kindt (1975)

Rabbit allotypic determinants are expressed by a small letter designating the group (or locus) followed by a number for the allele. The original proposal for the allotype nomenclature (Dray *et al.*, 1962) included an A before every notation (for example, Aa1, Ab4), but it is usual to omit this. A unique number is assigned to each allotype. Phenotypes are written with an oblique stroke separating the groups, such as a1a2/b4/d11d12. Genotypes are written with the alleles as superscripts, for example, $a^1a^2/b^4b^4/d^{11}d^{12}$. Allotypes not assigned to a group are written with an A preceding the number. For instance, the early reports of allotypes d11 and d12 would have referred to them as A11 and A12.

2.1.3 Structural Correlates of Allotypes

The validity of allotypes as genetic markers rests on the supposition that they are primary gene products. That is to say, the differences that the allotypic antisera recognize must reside in the primary structure of the immunoglobulin. Correlation of primary structure with allotypy is rather more complicated than a superficial examination of the problem would indicate (Todd, 1972). Detection of structural differences in allotypically different proteins is not proof that these are the differences which give rise to allotypic determinant. Furthermore, several kinds of variations other than allotypic variations have been described for antibody chains, including class, subclass, subgroup, and the binding site-associated hypervariable regions. Each type of variation must, of course, be recognized and taken into consideration. Therefore, to assign accurately allotypic correlates, a considerable background information on the general structural features of immunoglobulin chains in question is needed.

Amino acid substitutions that are present on chains of one allotype and that are different on chains of a second allotype are called correlates. Todd (1972) has pointed out that a correlate need not be involved in the antigenic determinant of the allotype. Proof that a given substitution in the amino acid sequence is indeed an allotypic determinant requires the isolation of homologous peptides and the subsequent demonstration that one peptide has one allotype and the second another. Such proof has been obtained in very few instances for rabbit allotypes (McBurnett and Mandy, 1974).

2.1.4 V_H Allotypes

The rabbit is unique in having allotypic markers present in the variable region of the heavy chain. Of these, the group a markers are by far the most thoroughly studied. While sequential studies of rabbit heavy chains are complicated by the fact that most of the chains are blocked at the N-terminus, sequence differences between group a allotypically defined pooled heavy chains have been demonstrated, as shown in Figure 2. The substitutions are extensive and in some instances, most notable in the peptide from residues 69–73, two of the allotypes are identical and the third different. This situation would be expected if allotypes represent mutants of a common ancestral gene (Kindt, 1975). Cross-reactions have been observed (Brezin and Cazenave, 1975) between a1 and a3 heavy chains which were recognized only by antisera prepared in a^2a^2 rabbits. These cross-reactions could be explained possibly by the type of substitutions observed at positions 69–73 in the heavy chains.

In certain cases, as shown in Figure 2, heavy chains from homogeneous antibodies have been observed to carry substitutions other than those observed for the pooled heavy chains (Jaton et al., 1973). Similarly, it has been shown that group a allotypic specificities expressed by homogeneous antibodies are immunologically deficient with respect to those of IgG pools (Kindt et al., 1973c). These findings suggest that group a allotypy may comprise a group of subspecificities in the V_H region.

That certain heavy chains lack group a allotypes has been recognized for some time (Dray and Nisonoff, 1963). In homozygous allotype suppression experiments, David and Todd (1969) were able to raise the level of group a-negative IgG

POSITION

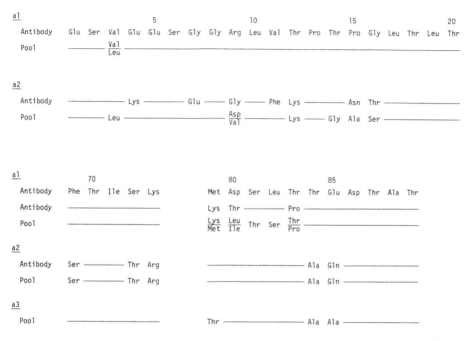

Figure 2 Sequences from V_H regions showing variation correlated with group a allotype. The regions are 1–20, 69–73, and 79–89. The sources of the sequences shown are pools (Mole *et al.*, 1971; Porter, 1974; Mole, personal communication); a1 antibodies (Jaton and Braun, 1972; Jaton *et al.*, 1973; Jaton, 1974*a*); and a2 antibody (Fleischman, 1971, 1973; Jaton, 1975)

molecules to nearly 100 per cent. Also, in several instances, homogeneous antibodies produced by streptococcal injection have been shown to lack group a allotypes (Kindt *et al.*, 1970*b*, 1973*a*). The a-negative heavy chains have been shown to have distinct structural features, including amino acid composition (Tack *et al.*, 1973) and N-terminal sequence. Amino acid composition differences were shown to be limited to the N-terminal half of the heavy chain. The distribution of N-terminal peptides recovered from a1, a2, a3, and three a-negative (ā) preparations derived by allotype suppression in a homozygote of each allotype is shown in Table 2.

Table 2 Distribution of N-terminal peptides in different allotypes

Peptide	Yield (moles of peptide/mole of heavy chain)					
	a1	a2	a3	ā(a1)	ā(a2)	ā(a3)
pGlu-Ser-Val-Glu	0·67	0·38	0·16	0·05	0·15	0·03
pGlu-Ser-Leu-Glu	0·06	0·34	0·63	0·03	0·09	0·06
pGlu-Glu-Glu	0·10	0·17	0·09	0·77	0·77	0·67

Distinct groups of allotypic markers also have been found on a-negative heavy chains (Kim and Dray, 1972, 1973). At least two allotypes in distinct groups (x32 and y33) are present on different populations of a-negative molecules. It should be emphasized that x32 and y33 are not allelic either to each other or to the allotypes a1, a2, or a3.

2.1.5 C_H Allotypes

The IgG C-region allotypes d11, d12, e14, and e15 are the only rabbit allotypes for which simple structural correlates are known. The group d allotypes were shown to correlate with methionine (d11)—threonine (d12) interchange at position 225 of the γ chain (Prahl et al., 1969). McBurnett and Mandy (1974) have demonstrated that a hinge region peptide having the methionine substitution has d11 antigenic activity.

Group e allotypy was shown also to correlate with a single amino acid interchange in the C_γ region (Appella et al., 1971). These allotypes were correlated with threonine (e14) and alanine (e15) at position 309. No other differences among tryptic peptides from the Fc fragments have been noted.

No structural correlates have been reported as yet for the allotypes of rabbit IgA or IgM.

2.1.6 C_L Allotypes

The group b allotypes, found in the C region of chains of both kappa subtypes, appear from preliminary sequential studies to correlate with multiple amino acid substitutions, like the group a allotypes. Unlike the group a allotypes, however, the serological specificities of the group b allotypes appear to be identical on homogeneous antibodies and among IgG pools (Kindt et al., 1972). The first allotypically different peptides reported for group b allotypes were at the C terminals of the light chains (Appella et al., 1969; Frangione, 1969; Goodfleisch, 1975) (Figure 3).

Figure 3 Light chain C-terminal peptide sequences of group b allotypes

b4 Asn- Arg- Gly- Asp- Cys
b5 Ser ——Lys- Asx ——
b6 Ser ——Lys-Ser ——
b9 ——————

Other significant differences have been reported among the b4, b5, and b9 peptides containing cysteine 171 (Lamm and Frangione, 1972; Chen et al., 1974; Goodfleisch, 1975). More recently, studies on b9 light chains in several laboratories have demonstrated extensive differences between b4 and b9 in positions 110–130 (Hood et al., 1975; Zeeuws and Strosberg, 1975; Thunberg, unpublished results). As shown in Figure 4, the b9 sequences differ from the b4 in at least seven positions, and they differ from one another in at least one position. In the authors' laboratory, a b4 light chain from a homogeneous anti-streptococcal antibody (4539) has been found which differs from the prototype b4 sequence at two positions, including 121 where it has the serine residue

L CHAIN SEQUENCES - POSITIONS 110–130

	110					115					120					125					130
b4 4135[a]	Asp	Pro	Val	[ ]	Ala	Pro	Thr	Val	Leu	Ile	Phe	Pro	Pro	Ala	Ala	Asp	Gln	Val	Ala	Thr	Gly	Thr Val
b4[b]	Gln	Ile							Leu				Ser					Leu				
b9[c]									Leu				Ser				Glx	Leu	Thr	Gly	Glx	
b9[d]	Xxx	Ile							Leu				Ser					Leu			Thr	Glu

Figure 4 C_L sequences of two b4 and two b9 chains in the region 110–130. [a] Chen et al. (1974); [b] Sogn, unpublished results; [c] Zeeuws and Strosberg (1975); [d] Hood et al. (1975) for b9 pool and Thunberg (unpublished results) obtained for a homogeneous antibody light chain

associated only with b9 (Sogn, unpublished results). Antibodies of the b4 allotype show considerable variation in sequence at position 174 (Strosberg *et al.*, 1972; Appella *et al.*, 1973). The complication of intra-allotype C-region sequential variation, combined with the extensive interallotype differences, makes identification of residues involved in the allotypic determinant as yet impossible.

Allotype suppression has been used to increase the relative concentration of λ chains in rabbits to levels which allow more detailed studies (Appella *et al.*, 1968; Rejnek *et al.*, 1969; Chersi and Mage, 1973). The two markers described for these chains, c7 and c21, (Mage *et al.*, 1968) have been shown to be pseudoalleles (Gilman-Sachs *et al.*, 1969). No structural correlates are available.

2.1.7 Genetic Markers for the V_L Region

Structural and serological evidence suggests that the V_L region is not involved in group b allotypic variation (Appella *et al.*, 1969; Frangione, 1969; Kindt *et al.*, 1972). While no V_L allotypes have been described, genetic variants detected by sequential analysis, which do not give rise to antigenic differences, are known. Studies of these sequence variations have shown them to be associated preferentially with different group b allotypes.

Studies by Waterfield *et al.* (1973) on the N-terminal sequences of allotypically defined light-chain pools indicate allotype-related variation of a quantitative nature, which is sufficiently reproducible to allow 'typing' of the b allotypes by this method. Furthermore, Thunberg *et al.* (1973) have detected a glutamate residue at position N16 of certain homogeneous b9 light chains, a substitution never previously reported for light chains of any species. This substitution was found in b9 light chain pools only obtained from rabbits descended from one of two b9 progenitor rabbits in the colony. Thus, the glutamate 16 substitution may represent an inherited difference in the rabbit V_L region.

Other substitutions, which occur in light chains of allotype b4, but not b9, have been reported at position 2 (tyrosine) and at positions 12 and 13 (glutamate). These substitutions occur in b4 light chains over three generations in one family, and in b4 IgG pools. In addition, they have been found in roughly the same frequency on a list of 41 homogeneous b4 light-chain N-terminal sequences as found in pooled IgG.

Since none of the substitutions described above is found in all chains of the allotype with which they are associated, they are unlikely to be allotype related. If they represent subgroup differences, then their allotype-related absences are difficult to explain. It seems more likely that these substitutions indicate the presence of a finite number of V_L genes linked to the C_L genes.

2.2 Allotypes of the Mouse

The mouse has several advantages over the rabbit for genetic studies. These include the obvious ones of smaller size and a shorter gestation period, as well as the availability of inbred and congenic strains. In addition, transplantable plasmacytomas, which can produce immunoglobulins in sufficient amounts for structural studies, are induced readily in the mouse (Potter, 1967). Immunogenetic studies in mice have been limited by the fact that allotypic markers have been identified only for the C_H regions. This shortcoming has been circumvented partially by the use of idiotypes as V_H markers in conjunction with the C_H allotypes to obtain information on organization of immunoglobulin genes in this species. Furthermore, a genetic marker for the V_H region of mouse chains, detectable by structural means, has been described (Edelman and Gottlieb, 1970).

Structural correlates for mouse allotypes have not been described, but the recent completion of the amino acid sequence of a murine IgG heavy chain (Bourgois et al., 1974) indicates that such data should be available in the near future.

2.2.1 Immunoglobulins of the Mouse

There are six heavy-chain immunoglobulin classes in the mouse. These are referred to by different designations. Most reports use one of the following notations

γM	IgM		
γA	IgA		
γ1	IgF	IgG1	
γ2a	IgG	IgG2a	
γ2b	IgH	IgG2b	
γ3	IgI	IgG3	

The general structural features of mouse immunoglobulins are similar to those in man. The light chains have two intra-chain disulphide bridges and the heavy chains have four. The light chains of the mouse may either be κ or λ. As with the rabbit, the former comprises the large majority (97 per cent) of the light chains in normal circulation. The number of V_L subgroups for mouse κ chains may be greater than 20 based on sequential studies utilizing κ chains from BALB/c plasmacytomas (Hood et al., 1973). Two λ V-region subgroups have been described (Mage et al., 1973).

2.2.2 Mouse C_H Allotypes

Allotypes have been described for the C region of mouse immunoglobulin classes F, G, H, and A. No allotypes have been described for classes IgI or IgM. In addition, there is a group of allotypes which have not yet been assigned to any heavy chain class. The early work on mouse allotypes has been reviewed by Potter and Lieberman (1967), and by Herzenberg et al. (1968). Mage et al. (1973) have presented more recently a thorough review.

The mouse allotypes have been named by two different systems. Each uses a designation for the heavy-chain class on which the allotype is expressed, followed

Table 3 Alignment of mouse allotypes reported by different investigators*

IgG		IgH		IgF		IgA	
1	2	1	2	1	2	1	2
G^1	1·10	H^9	3·4	F^{19}	4·1	A^{12}	2·2
G^3	1·3	H^{11}	3·2	F	4·2	A^{13}	2·3
G^5	1·11	H^{16}	3·9	F^8		A^{14}	2·4
G^4		H^4				A^{15}	
G^6	1·12	H^{22}				A^{17}	
G^8							
G^{3+8}							

1. The nomenclature of Potter and Lieberman (1967). 2. The nomenclature of Herzenberg *et al.* (1968). Only the allotypes 2 which have counterparts in the system 1 are listed here
* In addition to the determinants listed here, there are determinants 2, 10, 18, 21, and 24 which have not been assigned to a heavy-chain class

by a number to describe the determinant. In the system used by Potter and Lieberman (1967), the class designations are G, A, H, and F: that is, the same as the immunoglobulin classes. The determinants are then numbered in order of their discovery, using a unique number for each determinant. Therefore, the allotype G^1 is an IgG allotypic determinant, presumably the first observed by these workers. If the same number is used twice with different letters (for example, G^8 and F^8) this means that the same determinant has been detected on two different classes of immunoglobulin.

A second system (Herzenberg *et al.*, 1968) uses a number for the class (IgG = 1, IgA = 2, IgH = 3, and IgF = 4) and a second number, separated from the first by a full stop, for the determinant. The numbers are repeated for each set of determinants. Therefore, in this system, 1.1 and 2.1 are not the same, but rather would indicate different determinants on IgG and IgA, respectively.

Mage *et al.* (1973) published a table showing alignment of the allotype determinants as named by the different systems. This useful alignment is reproduced in a simpler form (Table 3) omitting allotypes for which the class has not been ascertained yet and omitting the determinants of Herzenberg *et al.* (1968), which do not have a corresponding determinant in the system of Potter and Lieberman.

The mouse allotypes are often referred to by the inbred strain in which they occur. For example, the complete BALB/c allotype ($G^{1,6,7,8}$, $A^{12,13,14}$, $H^{9,11,22}$, $F^{8,19}$) is too cumbersome therefore often it is called the BALB/c allotype. This refers to the entire C_H-gene complex present in the BALB/c strain. This notation is convenient when congenic strains are being discussed.

The determinants for a single class within a single strain may be designated also by a condensed notation using the Herzenberg number for class designation with a small letter as a superscript for the allele. The BALB/c IgG allotype ($G^{1,6,7,8}$) is written as 1^a and the C57BL/6(G^-) as 1^b. In some reports these may be written as

IgG^a and IgG^b, or even as G^a and G^b. A listing of alleles present in various inbred strains is given by Mage *et al.* (1973).

2.2.3 Genetic Marker for the Variable Region of Mouse κ Chains

Edelman and Gottlieb (1970) observed that a peptide detected by autoradiography from the V$_L$ region of certain mouse κ chains would serve as a genetic marker. This marker was found in three of 17 strains surveyed. More recently, Gottlieb (1974) has shown correlation between this marker, called I$_B$-peptide marker, and the thymocyte cell-surface antigen Ly-3.1. This finding, which represents the first report of linkage between an immunoglobulin gene and another characteristic, would place the putative V$_L$ gene in linkage group XI on chromosome 6 of the mouse (Itakura *et al.*, 1972).

3 IDIOTYPES AS VARIABLE-REGION MARKERS

The presence of serological markers in both C and V regions of an immunoglobulin chain enables studies to be made concerning the number and nature of the genes for antibody V regions and the relationships between these genes and those encoding C regions. C-region allotypes are quite readily available in both rabbits and mice, but the only V-region allotype described is the rabbit group a in the V$_H$ region. Structurally defined genetic markers, such as those described earlier for rabbit and mouse V$_L$ regions, offer an alternative to V-region allotypes. A second alternative is the use of idiotypes. Idiotypic phenomena have been reviewed recently by Capra and Kehoe (1975).

The individual antigenic specificity of human myeloma proteins was recognized initially by Slater *et al.* (1955), who used absorbed rabbit antibodies to detect the differences between myeloma proteins. Oudin and Michel (1963) observed that rabbit antibodies to bacterial vaccines also possessed very restricted antigenic determinants. The antisera of Oudin were prepared in allotypically matched rabbits (Oudin and Michel, 1969a, b). Oudin named this phenomenon *idiotypy*.

Idiotypic determinants are binding-site related and some idiotypic reactions may be inhibited by haptens against which the antibodies are raised (Brient and Nisonoff, 1970; Weigert *et al.*, 1974). Many idiotypes have been shown to require specific heavy–light chain combinations for expression (Grey *et al.*, 1965). The idiotypes of two human myeloma proteins with anti-IgG activity have been related recently to near identity of amino acid sequences in heavy-chain hypervariable regions (Capra and Kehoe, 1974). While it may be premature, it is tempting to define idiotypes as antigenic determinants related to hypervariable regions of immunoglobulins.

Use of idiotypes as markers became possible when studies on idiotypes of homogeneous antibodies in rabbit families (Eichmann and Kindt, 1971) established that idiotypes are genetically transmitted. Further work on the idiotypes of rabbit antibodies showed linkage of group a allotypes to idiotypes

(Kindt *et al.*, 1973*b*, *c*; Kindt and Krause, 1974). In addition, the association of an idiotype with an infrequently occurring V_κ subgroup has been demonstrated in these same rabbit families (Klapper and Kindt, 1974). Genetic experiments on the idiotypes of the mouse which have extended the original observations will be discussed in the following section.

4 RELATIONSHIPS BETWEEN IMMUNOGLOBULIN GENES

4.1 General Considerations

Investigations of immunoglobulin allotypes have placed the genes coding for immunoglobulin synthesis at three distinct loci. These loci code for the heavy chains, the κ light chains, and the λ chains. The original observation was made by Dubiski *et al.* (1962), who noted that the two groups of allotypic markers for heavy chains and κ light chains (Oudin, 1956, 1960*a*, *b*) showed independent assortment. Genetic studies on the group c allotypes of the λ chains have added the third unlinked group (Gilman-Sachs *et al.*, 1969). The recent observation of secretory-piece allotypes, which are not linked to any of the other allotypes, may add a fourth locus (Knight *et al.*, 1974*b*).

All the C_H and V_H allotypes so far observed for the rabbit heavy chain are linked to one another. Likewise, mouse C_H allotypes are linked to each other and to the idiotypes which have been used as V_H gene markers. Because serological markers for genetic variants in V_L regions are not available, there is no similar linkage information for the κ and λ light-chain loci. Data on structurally detected markers of the V region of the rabbit κ light chains (Thunberg *et al.*, 1973; Waterfield *et al.*, 1973; Thunberg, 1974) suggest that the light-chain gene complex consists of few C-region genes closely linked to larger numbers of V-region genes.

4.2 Heavy-Chain Gene Complexes

The availability of allotypic markers for several mouse heavy-chain classes has made it possible to test the linkage between the genes encoding the different C regions. In back-crosses involving 2371 progeny, no recombinant types were observed, indicating a very low recombination frequency among the genes which encode the C_H allotypes. The existence of allotypic combinations in wild mice which do not occur in inbred strains suggests that crossovers can and have occurred among these closely linked genes (Lieberman and Potter, 1969).

Data obtained on the linkage of rabbit allotypes in the C_H region concur with these conclusions. That is, in the rabbit as well as in the mouse, the C_H genes are very closely linked. Again, the existence of various combinations of the rabbit C_H genes indicates that at some time crossovers have occurred among them. The presence of a limited number of possible linkage groups, called 'allogroups', has been observed in both species. For example, the rabbit IgA allotypic groups f and g have five and four alleles, respectively, but of the twenty possible combinations,

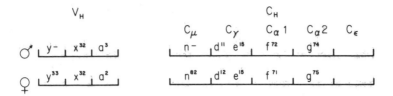

Figure 5 The complete heavy-chain typing of a single rabbit. The allogroup inherited from the paternal side is at the top and that from the maternal side is at the bottom

only five have so far been observed (Hanly and Knight, personal communication).

A striking example of restriction is obtained by considering the entire heavy-chain allogroup of the rabbit (Figure 5), with its three V_H-region (a, x, and y) and five C_H-region (d, e, f, g, and n) allotypic groups. The total number of possible combinations may be estimated by multiplying the number of alleles in each group (see Table 1). In this manner, an estimate of 2880* different allogroups is obtained. Of these 2880 possibilities, only 12 have been observed (Mage *et al.*, 1973)—less than 0·5 per cent of the maximum number of combinations.

Such restriction of possible gene combinations which seems to be the rule for the rabbit and the mouse systems has also been observed with human immunoglobulin allotypes (Natvig and Kunkel, 1973). The reason for the existence of so few allogroups is obscure. It may be that certain combinations are deleterious and are, therefore, selected against. Alternatively, the linkage of the C_H genes may be so close that very few of the theoretically possible crossovers can occur in practice.

4.3 V_H and C_H Allotypes and Idiotypes

The genetic relationships between the allotypes of the C_H and V_H regions have implications beyond the field of immunology. The original observation of Todd (1963) that IgM and IgG have in common a V-region allotype suggests that two or more genes interacted prior to the synthesis of a single polypeptide chain. The presence of these V-region allotypes was demonstrated subsequently for rabbit IgA (Feinstein, 1963) and IgE (Kindt and Todd, 1969). The observation of identical idiotypes on human myeloma proteins with different heavy-chain classes (Penn *et al.*, 1970; Wang *et al.*, 1970) has provided strong independent confirmation of the 'two gene—one polypeptide' hypothesis. Also pertinent to the relationships of V_H and C_H genes is the observation of Fu *et al.* (1975), who have recently shown that the IgM and IgD molecules on the surface of human lymphocytes have the same idiotypic determinants. This simultaneous presence on one cell of two different chains with identical V regions must be taken into

---

* This is a conservative estimate because the number of alleles in several groups is not known and only one additional allotype has been allowed for these. Furthermore, the existence of subclasses and allotypes of other classes not yet well described in the rabbit (IgE, IgD) would raise this number.

account in molecular mechanisms which seek to explain the production of antibody heavy chains by translocation events.

The genetic relationships between V and C-region genes were tested experimentally with the discovery of rabbit heavy-chain C-region allotypes (Mandy and Todd, 1968; Dubiski, 1969a). It was soon shown (Zullo et al., 1968; Dubiski, 1969b) that the C-region allotypes of group e and group d are linked to the V-region allotypes of group a. Subsequently, linkage of idiotypes to C_H allotypes has been demonstrated in the mouse for a large number of idiotypes. Crossovers between the C and V-region allotypes have been documented for rabbit families (Mage et al., 1971; Kindt and Mandy, 1972; Hamers-Casterman and Hamers, 1975), thus providing further evidence for the involvement of two separate genes in antibody heavy-chain synthesis. The crossover frequency for the C_H and V_H genes of the rabbit based on these observed recombinations has been estimated to be 0·3 per cent (Mage et al., 1973). Low recombination frequencies also have been reported for mouse idiotypes and C_H allotypes (Blomberg et al., 1972; Eichmann and Berek, 1973; Pawlak et al., 1973; Lieberman et al., 1974).

By studying combinations of V and C allotypes present on the molecules from rabbits doubly heterozygous for these markers, it was observed that a high percentage of the heavy chains retains the combinations of markers inherited from the parents (Kindt et al., 1970a). That is, the majority of syntheses are directed by genes in the coupling phase. In the example given in Figure 6 the majority of molecules in the offspring's circulation with allotype a1 would express d11; those with a2 would express d12. Similarly, in the mouse, Eichmann (1973) showed that idiotypically related antibodies produced in allotypically heterozygous mice always have the same allotype–idiotype combination. This indicates that usually mouse idiotypes, like rabbit allotype, are synthesized from

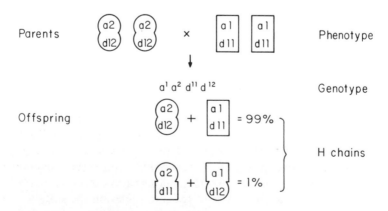

Figure 6 An illustration of the preference in a rabbit for heavy-chain synthesis using V and C genes inherited on the same chromosomes. Parental chromosomes (or heavy chains) are shown at the top. In the offspring, 99 per cent of circulating heavy chains have the same allotypic combination as one or the other of the parents

information present on the same chromosome as the C_H gene with which they are transmitted genetically.

In rabbits, the estimates of molecules with recombinant types obtained by the selective absorption experiments of Landucci-Tosi and Tosi (1973) agree well with the estimate of 1 per cent which Pernis *et al.* (1973) obtained using double fluorescent staining techniques. More recent experiments have measured the percentage of recombinant molecules utilizing IgA C-region allotypes in combination with the allotypes of group a. The numbers obtained are slightly higher than those for IgG (Knight *et al.*, 1974a). It is not known whether the recombinant molecules result from synthesis directed by genes in repulsion or by genes on chromosomes which result from mitotic crossovers between C_H and V_H genes.

The recognition of a number of idiotypes in inbred mouse strains makes it possible, by examining recombinants among the idiotypes and the allotypes to which they are linked, to begin to map the area of the chromosome where these genes are found and thereby to estimate the number of V_H genes present. For example, one study involved the linkage between two different mouse idiotypes (A5A and ARS), both linked to the A/J C_H allotype (Eichmann *et al.*, 1974). The observation of a crossover between the A5A idiotype and the BALB/c allotype in one male mouse provided the opportunity to test the linkage between the two idiotypes. The crossover mouse was bred and the new haplotype (allogroup) was shown to be inherited without the ARS idiotype. On the basis of this data, the chromosomes of the BALB/c, A/J and recombinant might be as represented as shown in Table 4.

Table 4   Crossing over in V-region genes between BALB/c and A/J mice

	V_H Idiotypes		C_H Allotypes
A/J	A5A$^+$	ARS$^+$	1.e
BALB/c	A5A$^-$	ARS$^-$	1.a
Recombinant	A5A$^+$	ARS$^-$	1.a

Such an observation would suggest that V-region genes are present in sufficient numbers to observe crossovers among them. Whether the ARS and A5A idiotypes occur in heavy chains with the same or different V_H subgroups is not known. A more complete survey of mapping data in the mouse has been presented by Eichmann (1975).

4.4 Interaction of Multiple Genes in the Synthesis of Single Immunoglobulin Variable Regions

As discussed in Section 4.3, the synthesis of a single immunoglobulin chain is probably directed by more than one gene. The existence of separate V and C genes has been most firmly established by allotypic and structural studies of heavy chains (Todd, 1963; Dreyer and Bennett, 1965). More recently, structural and serological data on V regions have led to the further suggestion that V-region synthesis may be directed by multiple genes. The experiments supporting this hypothesis have been discussed by Capra and Kindt (1975), who suggested that three types of genes interact in the synthesis of a single antibody chain: (*i*) C-region genes which specify isotypes; (*ii*) the V genes which specify the relatively invariant, or framework residues of the V region, including subgroup-specific residues and any allotypic determinants; and (*iii*) the V genes which specify the hypervariable regions and the idiotype.

Structural studies on the V regions of antibodies in myeloma proteins indicate that hypervariable regions are interspersed between V-region sequences which show very little or even no variation among immunoglobulins from different individuals or distinct species (Wu and Kabat, 1970; Capra and Kehoe, 1975). Because of the problem of genetic load (Ohno, 1970), the existence of these invariant portions may be explained best by theories based on few genes, while structural studies showing conservation of hypervariable regions in genetically dissimilar individuals may best be explained by multigene theories. Gene-interaction theories explain this constancy and variability within the same polypeptide chain by the existence of multiple gene segments which interact to synthesize a single V region.

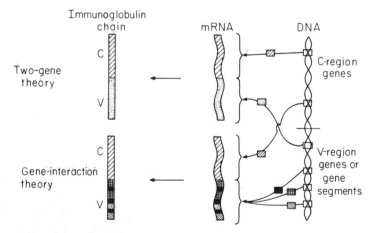

Figure 7 An example of a gene interaction theory, as compared to a two-gene theory. In the two-gene theory, one C-region gene and one V-region gene together code for a mRNA specifying a complete immunoglobulin chain. In the gene-interaction theory, multiple V-region gene segments interact with one C-region gene to yield the mRNA. In the example, three types of V-region gene segments are suggested to be involved, one for the N-terminal half of the V-region, one for the C-terminal half and one for the hypervariable sections. Other formulations are possible within the framework of the gene-interaction hypothesis

Results supporting gene-interaction theories have been obtained in serological studies on combinations of V_H allotypes and idiotypes (Kindt *et al.*, 1973*a*). It has been shown that two antibodies with identical idiotypes differ in their V-region allotypes and N-terminal sequences (Waterfield *et al.*, 1972). Structural studies on myeloma proteins with shared idiotypic specificities similarly have supported the interaction theory (Kunkel *et al.*, 1973). Sequential analysis of the V_H regions of two idiotypically similar myeloma proteins from non-related individuals indicates differences in eight positions. Only three of these differences occur in the hypervariable regions (Capra and Kehoe, 1974). Randomly selected myeloma proteins from the same V-region subgroup will differ by an average of 33 of the 42 hypervariable positions. More recently, studies on localization of variable residues within the framework residues of the V region have shown that variability associated with subgroup or species is confined to the N-terminal portions of this region. Positions toward the C-terminal end of the V region show no such pattern. Such an observation argues that these regions may be encoded by separate genes.

There is no direct evidence for gene interaction or for any other theory of V-region diversity. There is a precedent, however, for gene interaction from studies on V and C-region genes. The upper portion of Figure 7 depicts the synthesis of a single immunoglobulin chain from separate V and C-region genes. In the lower portion of Figure 7 this well-accepted event is extended to include interaction between V-region gene segments. While mechanisms for the proposed interactions are conjectural, there is evidence to suggest that translocation events occur prior to translation (Capra and Kindt, 1975).

5 SELECTIVE EXPRESSION OF ALLOTYPE GENES

Although individual plasma cells may produce and simultaneously express immunoglobulins of several classes on their membranes (Warner, 1974), in allotypic heterozygotes a single plasma cell uses only one of the two allotypes available in each allotypic group in its genome. Thus, immunoglobulin-producing cells may be said to exhibit a mosaic of phenotypes. Because the genes coding for heavy and light chains are unlinked at least two different loci, perhaps on different chromosomes, must be involved. This phenomenon, generally referred to as allelic exclusion, was first proposed on the basis of allelic restriction observed for myeloma proteins (Harboe *et al.*, 1962*b*). This was demonstrated subsequently at the cellular level for rabbit allotypes by Pernis *et al.* (1965) and for mouse allotypes by Weiler (1965). The term allotype selection will be used here in preference to allelic exclusion as recent data, to be discussed in Section 6, indicate a strong possibility that at least some allotypic groups are not composed of true alleles. Regardless of the nomenclature, the phenomenon, as a unique example of co-ordinated selective gene activation or inactivation at multiple loci, is an important component of the mechanism regulating the immune response.

While an individual antibody-producing cell in the rabbit exhibits only one allele of each group, any combination of paternal or maternal allotypes may be observed. Because the allotypes of group a and b are unlinked and presumed to be on different chromosomes, this situation necessitates the activation, or inactivation, of genes at two distinct loci. If the allotypes of the λ light chains and the secretory component are taken into consideration, then the activation or inactivation process must take place in an antibody-producing cell at four distinct loci, perhaps on four different autosomes. The fact that the synthesis of 99 per cent of heavy chains utilizes information from genes in coupling also must be taken into account (see Figure 6). The gene-activation process may then have specificity for genes on the same chromosome (Tosi *et al.*, 1974). If this were the case, the low percentage of molecules expressing non-linked markers would be synthesized by cells in which somatic crossovers have occurred.

Another example of selective gene expression, and one which has points in common with allotype selection, is the utilization of only one X-chromosome in somatic cells of female mammals. A female with two X chromosomes, at a certain developmental state, randomly activates (or inactivates) one of the two chromosomes in each cell. When this was first postulated by Lyon (1961), the major evidence cited was the occurrence of the 'mottled' or 'dappled' coat colour phenotypes in heterozygous female mice and the existence of normal XO females. This has been confirmed in many subsequent studies using a number of X-linked traits, some of which can be assayed at the level of single cells. For example, Beutler *et al.* (1962) demonstrated that single erythrocytes from human females heterozygous for glucose 6-phosphate dehydrogenase deficiency exhibit either the normal or the deficient phenotype.

In both cases of selective gene expression, certain conclusions may be drawn. Firstly, it is likely that the process involves activation rather than inactivation. Individuals with more than the normal complement of X chromosomes have only one which is activated. It is much simpler to postulate a mechanism for the selective activation of one set of immunoglobulin genes than for inactivation of the many not used. Secondly, it is a random event in that either X chromosome may be activated and any of the many possible immunoglobulins may be synthesized. Immunoglobulin-gene activation may not be completely random, since studies have shown an imbalance of one allotype in pre-immune serum for allotypically heterozygous animals (Kindt, 1975). One of the possible explanations for these findings is the preferential activation of one locus. Thirdly, it is a permanent event. The progeny cells are committed to maintain their specific state of activation throughout all successive cell divisions and differentiative processes. It is reasonable to postulate because of these similarities that the same mechanism underlies both the autosomal and X-linked phenomena.

No mechanism yet proposed for either case of selective gene expression has received any significant experimental support. However, the model recently proposed by Riggs (1975) for selective activation of X chromosomes is particularly promising, but requires experimental verification.

In this model, Riggs has proposed that DNA methylation is the mechanism for X-chromosome activation. Enzymes, which are present in *Escherichia coli*, have been shown to methylate DNA forming N-6-methyladenine or 5-methylcytosine.

The methylation rate for DNA that has the opposing strand methylated is 10^3 times faster than the rate for strands which are not methylated. The methylated bases do not interfere, however, with pairing of DNA strands. Riggs has postulated that the methylation of an inactivation centre (a specifically recognized base sequence) on one X chromosome renders this chromosome active. In all subsequent cell divisions, the same centre would be rapidly methylated because one strand is already methylated. Thus, randomness and permanence are explained easily by this model. Prevention of methylation during oogenesis or spermatogenesis would allow formation of the derivatives to be reversed.

Riggs' model also may be applied to autosomal activation phenomena with minor changes. The major difference lies in the necessity for inactivation centres at several loci that may serve to inactivate only portions of the chromosomes on which they reside. A positive aspect of the Riggs model for selective gene activation is that it can be tested by the measurement of methylated bases in DNAs from various cells and by determining the presence of specific methylating enzymes.

6 NATURE OF ALLOTYPE GENES

It is assumed tacitly in many immunogenetic arguments that allotype and idiotype synthesis is controlled by structural genes. There is, however, some recent evidence suggesting that this may not be the case. One alternative might be that regulator genes exist and that these are inherited as autosomal codominant alleles. In this case, the information for all allotypic variants would reside in each individual and the expression of a set of allotypes would depend on the inheritance of the appropriate regulator genes. The mode of action of the regulator gene could involve specific activation or repression of the structural genes encoding the immunoglobulins.

Data consistent with such a model have come from the studies of Bell and Dray (1971), who showed that RNA extracts from rabbit spleen can induce synthesis of the donor's allotype in spleen cells from genetically different rabbits. Similarly, Rivat et al. (1973) showed that non-specific stimulation of cultured human lymphocytes gives rise to allotypes which have not been observed in the individual cell donors.

Members of a congenic mouse strain having the C57BL/ka allotype on a BALB/c background have been shown recently by Bosma and Bosma (1974) to express the BALB/c (IgG^a) allotype in a transient and unpredictable manner. These animals had been selected for homozygosity for the C57BL/ka allotype. Only the BALB/c allotype for immunoglobulins of the IgG classes were expressed. No other BALB/c allotypes were observed in these animals. Because the appearance of the so-called 'hidden allotype' in the homozygous animals appeared to coincide with general debilitation, it is considered possible that a virus causes the deviation from allelic behaviour. An alternative suggestion

134

invokes the presence of regulator genes which somehow became derepressed in the homozygotes and allow synthesis of the unexpected BALB/c allotype.

Data obtained from studies on rabbit allotypes also may be interpreted in terms of gene regulation. Strosberg *et al.* (1974) observed recently that a rabbit with allotype a^1a^3/b^4b^5 produced antibodies with allotypes a2 and b6 after immunization with *Micrococcus lysodeikticus*. The a2 and b6 allotypes were not necessarily on the same antibodies since the time of their appearances did not coincide. In addition to the rabbit producing these antibodies, two other rabbits in the Strosberg colony were found simultaneously to express three group a allotypes. The a2 allotype from the a^1a^3 was shown to be immunologically identical to normal a2 IgG using Ouchterlony analysis and by inhibition of radiobinding assays.

In the authors' colony, an a^1a^2 rabbit produced antibody of allotype a3 upon immunization with streptococcal C vaccine. Quantitative allotypic determination carried out on non-immune sera from a large number of rabbits representing different populations detected low levels of group a allotypes not identified by qualitative typing or anticipated from breeding data (latent allotypes) in 50 per cent of sera tested (Mudgett *et al.*, 1975). Allotype inhibition curves given by dilutions of a non-immune serum from an adult rabbit with nominal allotype a^2a^2 are depicted in Figure 8. This serum shows an appreciable concentration of IgG with allotype a3 and to a lesser extent a1. The latent allotypes, which are serologically identical to allotypes of pooled IgG, have been detected in sera from rabbits with all possible combinations of group a allotypes. Expression of latent allotypes in individual rabbits is transitory and sporadic.

Although the data of Strosberg *et al.* (1974) and the related findings in the authors' rabbits can be explained readily by failures in the postulated regulatory mechanisms, other explanations must be considered also. Among these is the

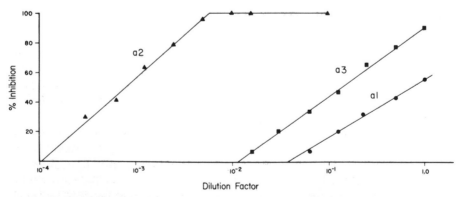

Figure 8 Inhibition of binding of 0·25 μg of radioiodinated a1b4, a2b4, or a3b4 IgG to *N*-hydroxysuccinimide-activated Sepharose immobilized anti-a1, anti-a2, or anti-a3, respectively, by dilutions of non-immune serum from rabbit 4120 (nominal allotype a2). Carrier proteins used in these assays include an excess of immunoglobulins with all a and b allotypes except that being measured. From Mudgett *et al.* (1975)

possibility that crossovers between V_H genes give rise to chromosomes simultaneously carrying genes for two different allotypes. Such an event would be more likely if there were many V_H genes in the germ line carrying the group a allotypes. Failure to observe rabbits with three group a allotypes has been used as an argument against the existence of multiple V_H genes in the germ line (Todd, 1972).

Whatever the exact nature of the allotype genes, the relationships between them as described in the previous sections must be taken into consideration. That is to say, a complete theory of immunoglobulin synthesis must take into account (*i*) the possible existence of separate V_H and C_H genes, (*ii*) the associations of V_H and C_H markers in the molecules of heterozygotes, and (*iii*) allelic exclusion of allotypes. Explanation of these phenomena must involve novel mechanisms, whether the postulated genes are structural or regulatory in nature.

7 CONCLUSION

The goal of the genetics of immunoglobulin is to determine the number and nature of the genes that interact and code for immunoglobulins, the arrangement of these genes in the genome, and the mechanisms behind the regulatory processes which determine what immunoglobulin will be produced by a single plasma cell. All of these remain current problems, but studies to date put significant constraints on any proposed model.

It is known that the genes which code for the allotypes of heavy chains, and κ and λ light chains are not linked, placing immunoglobulin genes at a minimum of three distinct loci. Genes encoding different C_H-region allotypes are linked closely and inherited as gene complexes called allogroups. Only a limited number of the many possible combinations of C_H genes has been observed in allogroups. There is good evidence that separate but tightly linked genes code for the V and C regions of a single immunoglobulin chain.

The number and nature of the V and C-region genes are unclear. All estimates indicate one, or very few C-genes, exist for each C-region allotype, but recent descriptions of polymorphisms within allotypically identical C regions must be taken into account. Competing theories suggest the presence of either few or many V-region genes for each group. Alternatively, there is appreciable evidence which suggests that V regions themselves may be the product of interaction between two or more distinct gene segments.

Evidence shows that about 99 per cent of γ and α heavy chains are synthesized from C_H and V_H genes interacting in coupling phase. This overwhelming preference for intra-chromosomal interaction puts constraints on any mechanism proposed for gene interaction. Additional constraints are imposed by the finding that individual cells can display simultaneously membrane IgM and IgD with the same idiotype. This implies that the same V_H gene may be utilized with two different C_H genes to synthesize these heavy chains.

Allotypically heterozygous single plasma cells express only one of the alternative allotypes available within each allotype group. This 'allotype exclusion' is required if clonal expansion is to lead to a specific response to antigenic expansion. The mechanism underlying this example of selective gene activation is unknown but intriguing hypotheses have been proposed. Description of regulatory mechanisms in immunoglobulin synthesis is further complicated by data which suggest that allotypes and hence immunoglobulin structural genes are not allelic. Since allotypes are inherited generally as simple Mendelian codominant alleles, it has been suggested that allelism is at the level of regulator genes and not structural genes. The unexplained complexities of immunoglobulin synthesis are such that it would certainly not be surprising to discover several levels of as yet undescribed regulator control over this vital physiological function.

ACKNOWLEDGEMENTS

This work was supported by PHS grants from the NIAID, AI 11995 and AI 11439. J.A.S. is a fellow of the Arthritis Foundation.

REFERENCES

Appella, E., Chersi, A., Mage, R. G., and Dubiski, S. (1971). *Proc. Nat. Acad. Sci., U.S.A.*, **68**, 1341.

Appella, E., Mage, R. G., Dubiski, S., and Reisfeld, R. A. (1968). *Proc. Nat. Acad. Sci. U.S.A.*, **60**, 975.

Appella, E., Rejnek, J., and Reisfeld, R. A. (1969). *J. Mol. Biol.*, **41**, 473.

Appella, E., Roholt, O. A., Chersi, A., Radzimski, G., and Pressman, D. (1973). *Biochem. Biophys. Res. Commun.*, **53**, 1122.

Bazin, H., Beckers, A., Deckers, C., and Moriame, M. (1973). *J. Nat. Cancer Inst.*, **51**, 1359.

Bazin, H., Beckers, A., and Querinjean, P. (1974a). *Eur. J. Immunol.*, **4**, 44.

Bazin, H., Beckers, A., Vaerman, J. P., and Heremans, J. F. (1974b). *J. Immunol.*, **112**, 1035.

Bell, C., and Dray, S. (1971). *Science, N.Y.*, **171**, 199.

Beutler, E., Yeh, M., and Fairbanks, V. F. (1962). *Proc. Nat. Acad. Sci., U.S.A.*, **48**, 9.

Blomberg, B., Geckeler, W. R., and Wiegert, M. (1972). *Science, N.Y.*, **177**, 178.

Bosma, M. J., and Bosma, G. C. (1974). *J. Exp. Med.*, **139**, 512.

Bourgois, A., Fougereau, M., and Rocca-Serra, J. (1974). *Eur. J. Biochem.*, **43**, 423.

Brezin, C., and Cazenave, P. A. (1975). *Immunochemistry*, **12**, 241.

Brient, B. W., and Nisonoff, A. (1970). *J. Exp. Med.*, **132**, 951.

Capra, J. D., and Kehoe, J. M. (1974). *Proc. Nat. Acad. Sci., U.S.A.*, **71**, 4032.

Capra, J. D., and Kehoe, J. M. (1975). *Adv. Immunol.*, **20**, 1.

Capra, J. D., and Kindt, T. J. (1975). *Immunogenetics*, **1**, 417.

Chai, C. K. (1974). *Immunogenetics*, **1**, 126.

Chen, K. C. S., Kindt, T. J., and Krause, R. M. (1974). *Proc. Nat. Acad. Sci., U.S.A.*, **71**, 1995.

Chersi, A., and Mage, R. G. (1973). *Immunochemistry*, **10**, 277.

Cohen, C., and Tissot, R. G. (1974). *Transplantation*, **18**, 150.

David, G. S., and Todd, C. W. (1969). *Proc. Nat. Acad. Sci., U.S.A.*, **62**, 860.

deVries, G. M., Lanckman, M., and Hamers, R. (1969). *Eur. J. Biochem.*, **11**, 370.
Dray, S., Dubiski, S., Kelus, A., Lennox, E. S., and Oudin, J. (1962). *Nature, (Lond.)*, **195**, 785.
Dray, S., and Nisonoff, A. (1963). *Proc. Soc. Exp. Biol. Med.*, **113**, 20.
Dreyer, W. J., and Bennett, J. C. (1965). *Proc. Nat. Acad. Sci., U.S.A.*, **54**, 864.
Dubiski, S. (1969a). *J. Immunol.*, **103**, 120.
Dubiski, S. (1969b). *Protides Biol. Fluids Proc. Colloq.*, **17**, 117.
Dubiski, S., Rapacz, J., and Dubiska, A. (1962). *Acta Genet.*, **12**, 136.

Edelman, G. M., and Gottlieb, P. D. (1970). *Proc. Nat. Acad. Sci., U.S.A.*, **67**, 1192.
Eichmann, K. (1973). *J. Exp. Med.*, **137**, 603.
Eichmann, K. (1975). *Immunogenetics*, **2**, 491.
Eichmann, K., and Berek, C. (1973). *Eur. J. Immunol.*, **3**, 599.
Eichmann, K., and Kindt, T. J. (1971). *J. Exp. Med.*, **134**, 532.
Eichmann, K., Tung, A., and Nisonoff, A. (1974). *Nature, (Lond.)*, **250**, 509.

Feinstein, A. (1963). *Nature, (Lond.)*, **199**, 1197.
Fleischman, J. B. (1971). *Biochemistry*, **10**, 2753.
Fleischman, J. B. (1973). *Immunochemistry*, **10**, 401.
Florent, G., deVries, G. M., and Hamers, R. (1973). *Immunochemistry*, **10**, 425.
Frangione, B. (1969). *FEBS Letters*, **3**, 341.
Franklin, E. C., Fudenberg, H., Meltzer, M., and Stanworth, D. R. (1962). *Proc. Nat. Acad. Sci., U.S.A.*, **48**, 914.
Fu, S. M., Winchester, R. J., and Kunkel, H. G. (1975). *J. Immunol.*, **114**, 250.

Gilman-Sachs, A., Mage, R. G., Young, G. O., Alexander, C., and Dray, S. (1969). *J. Immunol.*, **103**, 1159.
Goodfleisch, R. M. (1975). *J. Immunol.*, **114**, 910.
Gottlieb, P. D. (1974). *J. Exp. Med.*, **140**, 1432.
Grey, H. M., Mannik, M., and Kunkel, H. G. (1965). *J. Exp. Med.*, **121**, 561.

Hamers-Casterman, C., and Hamers, R. (1975). *Archs Int. Phys. Biochim.*, **83**, 188.
Hanly, W. C., Lichter, E. A., Dray, S., and Knight, K. L. (1973). *Biochemistry*, **12**, 733.
Harboe, M., Osterland, C., and Kunkel, H. G. (1962a). *Science, N.Y.*, **136**, 979.
Harboe, M., Osterland, C., Mannik, M., and Kunkel, H. G. (1962b). *J. Exp. Med.*, **116**, 719.
Herzenberg, L. A., McDevitt, H., and Herzenberg, L. A. (1968). *Ann. Rev. Genet.*, **2**, 209.
Hilschmann, N., and Craig, L. C. (1965). *Proc. Nat. Acad. Sci., U.S.A.*, **53**, 1403.
Hood, L., Campbell, J. H., and Elgin, S. C. R. (1975). *Ann. Rev. Genet.*, **9**, 305.
Hood, L., Eichmann, K., Lackland, H., Krause, R. M., and Ohms, J. J. (1970). *Nature, (Lond.)*, **228**, 1040.
Hood, L., McKean, D., Farnsworth, V. and Potter, M. (1973). *Biochemistry*, **12**, 741.

Itakura, K., Hutton, J. J., Boyse, E. A., and Old, L. J. (1972). *Transplantation*, **13**, 239.

Jaton, J. C. (1974a). *Biochem. J.*, **141**, 1.
Jaton, J. C. (1974b). *Biochem. J.*, **141**, 15.
Jaton, J. C. (1975). *Biochem. J.*, **147**, 235.
Jaton, J. C., and Braun, D. E. (1972). *Biochem. J.*, **130**, 539.
Jaton, J. C., Braun, D. G., Strosberg, A. D., Haber, E., and Morris, J. E. (1973). *J. Immunol.*, **111**, 1838.

Kim, B. S., and Dray, S. (1972). *Eur. J. Immunol.*, **2**, 509.
Kim, B. S., and Dray, S. (1973). *J. Immunol.*, **111**, 750.
Kindt, T. J. (1975). *Adv. Immunol.*, **21**, 35.

138

Kindt, T. J., Klapper, D. G., and Waterfield, M. D. (1973a). *J. Exp. Med.*, **137**, 636.
Kindt, T. J., and Krause, R. M. (1974). *Ann. Immunol. (Inst. Pasteur)*, **125C**, 369.
Kindt, T. J., and Mandy, W. J. (1972). *J. Immunol.*, **108**, 1110.
Kindt, T. J., Mandy, W. J., and Todd, C. W. (1970a). *Immunochemistry*, **7**, 457.
Kindt, T. J., Seide, R. K., Boksich, V. A., and Krause, R. M. (1973b). *J. Exp. Med.*, **138**, 522.
Kindt, T. J., Seide, R. K., Lackland, H., and Thunberg, A. L. (1972). *J. Immunol.*, **109**, 735.
Kindt, T. J., Seide, R. K., Tack, B. F., and Todd, C. W. (1973c). *J. Exp. Med.*, **138**, 33.
Kindt, T. J., Thunberg, A. L., Mudgett, M., and Klapper, D. E. (1974). In *The Immune System: Genes, Receptors, Signals*, (Sercarz, E. E., Williamson, A. R., and Cox, C. F., eds.), Academic Press, New York and London, p. 69.
Kindt, T. J., and Todd, C. W. (1969). *J. Exp. Med.*, **130**, 859.
Kindt, T. J., Todd, C. W., Eichmann, K., and Krause, R. M. (1970b). *J. Exp. Med.*, **131**, 343.
Klapper, D. G., and Kindt, T. J. (1974). *Scand. J. Immunol.*, **3**, 483.
Knight, K. L., Malek, T. R., and Hanly, W. C. (1974a). *Proc. Nat. Acad. Sci., U.S.A.*, **71**, 1169.
Knight, K. L., Rosenzweig, M., Lichter, E. A., and Hanly, W. C. (1974b). *J. Immunol.*, **112**, 877.
Krause, R. M. (1970). *Adv. Immunol.*, **12**, 1.
Kunkel, H. G., Agnello, V., Joslin, F. G., Winchester, R. J., and Capra, J. D. (1973). *J. Exp. Med.*, **137**, 331.

Lamm, M. E., and Frangione, B. (1972). *Biochem. J.*, **128**, 1357.
Landucci-Tosi, S. L., and Tosi, R. M. (1973). *Immunochemistry*, **10**, 65.
Lieberman, R., and Potter, M. (1969). *J. Exp. Med.*, **130**, 519.
Lieberman, R., Potter, M., Mushinski, E. B., Humphrey, W., and Rudikoff, S. (1974). *J. Exp. Med.*, **139**, 983.
Lyon, M. F. (1961). *Nature, (Lond.)*, **190**, 372.

Mage, R., Lieberman, R., Potter, M., and Terry, W. D. (1973). In *The Antigens*, (Sela, M., ed.), Academic Press, New York, p. 300.
Mage, R. G., Young, G. O., and Reisfeld, R. A. (1968). *J. Immunol.*, **101**, 617.
Mage, R. G., Young-Cooper, G. O., and Alexander, C. (1971). *Nature, New Biol.*, **230**, 63.
Mandy, W. J., and Todd, C. W. (1968). *Vox Sang.*, **14**, 264.
Margolies, M. N., Strosberg, A. D., Fraser, K. J., Perry, D. J., Brauer, A., and Haber, E. (1974). *Fed. Proc.*, **33**, 809.
McBurnette, S. K., and Mandy, W. J. (1974). *Immunochemistry*, **11**, 255.
Mole, L. E., Jackson, S. A., Porter, R. R., and Wilkinson, J. M. (1971). *Biochem. J.*, **124**, 301.
Mudgett, M., Fraser, B. A., and Kindt, T. J. (1975). *J. Exp. Med.*, **141**, 1448.

Natvig, J. B., and Kunkel, H. G. (1973). *Adv. Immunol.*, **16**, 1.
Nezlin, R. S., Vengerova, T. I., Rokhlin, O. V., and Machulla, H. K. G. (1974). *Immunochemistry*, **11**, 517.

O'Donnell, I. J., Frangione, B., and Porter, R. R. (1970). *Biochem. J.*, **116**, 261.
Ohno, S. (1970). *Evolution by Gene Duplication*, Springer Verlag, New York.
Oudin, J. (1956). *C. R. Acad. Sci., Paris*, **242**, 2606.
Oudin, J. (1960a). *J. Exp. Med.*, **112**, 107.
Oudin, J. (1960b). *J. Exp. Med.*, **112**, 125.

Oudin, J., and Michel, M. (1963). *C.R. Acad. Sci., Paris,* **257**, 805.
Oudin, J., and Michel, M. (1969a). *J. Exp. Med.,* **130**, 595.
Oudin, J., and Michel, M. (1969b). *J. Exp. Med.,* **130**, 619.

Pawlak, L. L., Mushinski, E. B., Nisonoff, A., and Potter, M. (1973). *J. Exp. Med.,* **137**, 22.
Penn, G. M., Kunkel, H. G., and Grey, H. M. (1970). *Proc. Soc. Exp. Biol. Med.,* **135**, 660.
Pernis, B., Chiappino, G., Kelus, A., and Gell, P. G. H. (1965). *J. Exp. Med.,* **122**, 853.
Pernis, B., Forni, L., Dubiski, S., Kelus, A. S., Mandy, W. J., and Todd, C. W. (1973). *Immunochemistry,* **10**, 281.
Porter, R. R. (1974). *Ann. Immunol. (Inst. Pasteur),* **125C**, 85.
Potter, M. (1967). *Meth. Cancer Res.,* **2**, 105.
Potter, M., and Lieberman, R. (1967). *Cold Spring Harbor Symp. Quant. Biol.,* **32**, 187.
Poulsen, K., Fraser, K. J., and Haber, E. (1972). *Proc. Nat. Acad. Sci., U.S.A.,* **69**, 2495.
Prahl, J. W., Mandy, W. J., and Todd, C. W. (1969). *Biochemistry,* **8**, 4935.

Reisfeld, R. A., Inman, J. K., Mage, R. G., and Appella, E. (1968). *Biochemistry,* **7**, 14.
Rejnek, J., Appella, E., Mage, R. G., and Reisfeld, R. A. (1969). *Biochemistry,* **8**, 2712.
Riggs, A. D. (1975). *Cytogenet. Cell Genet.,* **14**, 9.
Rivat, L., Gilbert, D., and Ropartz, C. (1973). *Immunology,* **24**, 1041.

Slater, R. J., Ward, S. M., and Kunkel, H. G. (1955). *J. Exp. Med.,* **101**, 85.
Strosberg, A. D., Fraser, K. J., Margolies, M. N., and Haber, E. (1972). *Biochemistry,* **11**, 4978.
Strosberg, A. D., Hamers-Casterman, C., van der Loo, W., and Hamers, R. (1974). *J. Immunol.,* **113**, 1313.

Tack, B. F., Feintuch, K., Todd, C. W., and Prahl, J. W. (1973). *Biochemistry,* **12**, 5172.
Thunberg, A. L. (1974). Doctoral Dissertation, The Rockefeller University, New York, N.Y.
Thunberg, A. L., Lackland, H., and Kindt, T. J. (1973). *J. Immunol.,* **111**, 1755.
Tissot, R. G., and Cohen, C. (1974). *Transplantation,* **18**, 142.
Todd, C. W. (1963). *Biochem. Biophys. Res. Commun.,* **11**, 170.
Todd, C. W. (1972). *Fed. Proc.,* **31**, 188.
Tosi, R. M., Landucci-Tosi, S., and Chersi, A. (1974). *J. Immunol.,* **113**, 876.

Wang, A. C. Wilson, S. K., Hopper, J. E., Fudenberg, H. H., and Nisonoff, A. (1970). *Proc. Nat. Acad. Sci., U.S.A.,* **66**, 337.
Warner, N. L. (1974). *Adv. Immunol.,* **19**, 67.
Waterfield, M. D., Morris, J. E., Hood, L. E., and Todd, C. W. (1973). *J. Immunol.,* **110**, 227.
Waterfield, M. D., Prahl, J. W., Hood, L. E., Kindt, T. J., and Krause, R. M. (1972). *Nature, New Biol.,* **240**, 215.
Weigert, M., Raschke, W. C., Carson, D., and Cohn, M. (1974). *J. Exp. Med.,* **139**, 137.
Weiler, E. (1965). *Proc. Nat. Acad. Sci., U.S.A.,* **54**, 1765.
Wu, T. T., and Kabat, E. A. (1970). *J. Exp. Med.,* **132**, 211.

Zeeuws, R., and Strosberg, A. D. (1975). *Archs Int. Physiol. Biochim.,* **83**, 41.
Zikan, J., Skarova, B., and Rejnek, J. (1967). *Folia Microbiol.,* **12**, 162.
Zullo, D. M., Todd, C. W., and Mandy, W. J. (1968). *Proc. Can. Fed. Biol. Soc.,* **11**, 111.

CHAPTER 5

Origin of Antibody Diversity

A. R. Williamson

1 INTRODUCTION

The origin of antibody diversity can be sought at three major levels: protein structure, cellular synthesis, and structural genes. This chapter sets forth the evidence for antibody diversity at each of these levels. Also, the various hypotheses that have been proposed to account for the antibody diversity are discussed.

The most basic question, which has been asked repeatedly, is whether the information for antibody synthesis precedes the challenge by antigen. This is the situation at the level of protein structure and at the level of cellular synthesis. At the genetic level, however, the question of whether information precedes antigenic challenge has not been answered completely. The uncertainty which remains concerns the proportion of the total pool of antibody structural genes expressed in a mature animal which is inherited in the germ line. Another way of posing this question is to ask to what extent the forces of mutation, recombination, and selection, which determine the diversity of the antibody gene pool, act either during evolution of the germ line or as somatic processes to amplify inherited diversity during the life of the individual.

2 STRUCTURAL BASIS OF ANTIBODY SPECIFICITY

2.1 Introduction

The best understood aspect of antibody diversity and specificity is the underlying structural basis. The fact that very similar immunoglobulin molecules can exhibit such a wide range of antibody specificities presented a paradox which has now been explained in terms of the amino acid sequences in each chain of each immunoglobulin molecule. It is accepted generally that the amino acid sequence of a protein determines its three-dimensional structure, and the specific demonstration of this fact for antibody molecules implies that a functional diversity of amino acid sequence is required to provide a set of different antibody combining sites. Antibodies of a different class have many physical and biological properties in common. This fact implies that considerable portions of the antibody molecule must have a common amino acid sequence. For each isotypic polypeptide chain the regions of constant amino acid sequence have been defined now. The region of amino acid sequence that varies from one immunoglobulin chain to the next has been the major focus of attention for those seeking to explain antibody diversity. Comparison of V-region amino acid sequences reveals a pattern of variability. Individual positions show a wide range of statistical variation. For each species, V-region sequences can be grouped into three distinct groups, corresponding to the κ, λ, and heavy chain C regions (Mage *et al.*, 1973). Within each set of V regions particular individual amino acids are invariant. The three-dimensional structure of two different immunoglobulin molecules indicates that a number of the constant amino acids are involved in making contact between the light and heavy chains (Davies and Padlan, 1975; Poljak, 1975). Other invariant or conservatively varied amino acids determine the characteristic folding for the peptide chain of each V region. This folding shows a great degree of similarity to that of each C-region domain.

2.2 Hypervariable Regions

The positions showing the highest degree of variability within each V region are found to be clustered together in three or four short portions which have been termed 'hypervariable regions' (Wu and Kabat, 1970; Capra and Kehoe, 1974*b*; Figure 1). The affinity-labelling data are provided in Table 1. The sequence in these hypervariable regions can be linked to the specificity of the antibody combining site.

(*i*) From these observations it is logical to deduce that a large proportion of the variation in sequence and V regions should contribute to functional diversity. Therefore, a role for the hypervariable regions in controlling antibody specificity might be expected.

(*ii*) Affinity-labelling studies using a variety of antibodies and myeloma proteins have identified many different positions in both V_L and V_H sequences as being close to or in the antibody combining site (Table 1). When these data are assembled together with a plot of variability, it can be seen that affinity-labelled amino acids or peptides lie in or adjacent to hypervariable regions (see Figure 1).

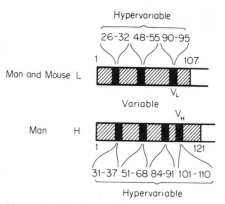

Figure 1 The location of some amino acids or peptides identified by affinity labelling relative to the hypervariable regions of V_H and V_L. (*a*)

In the case of protein MOPC 315 which binds DNP, a special affinity label having two active groupings was designed in order to show that the tyrosine 34 of the light chain and the lysine 54 of the heavy chain could be labelled simultaneously (Givol *et al.*, 1971). In this covalent joining of the two V regions, the distance between the tyrosine and lysine must be of the order of 0·5 nm.

(*iii*) In addition, correlation of the hypervariable-region sequence with antibody specificity has been shown in certain cases. Three examples are given here.

(*a*) *Human immunoglobulin M anti-γ globulins.* Some mixed cryoglobulins have been shown to consist of IgM antibodies specific for and complexed with IgG. From a group of 40 such IgM antibodies described by Williams *et al.* (1968), the

Table 1 Affinity-labelling data for location of some amino acids or peptides to the hypervariable regions of V_H and V_L

Species	Antibody specificity	Affinity-labelled positions	Reference
Light Chain			
Rabbit	DNP	Tyr 86	*a*
Pig	DNP	Tyr 33, Tyr 93	*b*
Mouse (TEPC15)	PC	Tyr 32	*c*
Mouse (MOPC315)	DNP	Tyr 34	*d*
Heavy Chain			
Rabbit	NAP	29–34, 95–114	*e*
Mouse (MOPC315)	DNP	Lys 54	*f*
Guinea-pig	DNP	Tyr 32, Tyr 60, Tyr (99–119)	*g*

a, Singer and Thorpe (1968); *b*, Franek (1971); *c*, Chesebro and Metzger (1972); *d*, Goetzl and Metzger (1972); *e*, Fisher and Press (1974); *f*, Givol *et al.* (1973); *g*, Ray and Cebra (1972). DNP, 2,4-dinitro phenyl; PC, phosphorylcholine; NAP, 2-nitro-4-azo-phenyl

Table 2 Structural differences between human non-selected $V_H III$ regions and V regions selected by cross-idiotypy

	Total sequence (124)	Hypervariable regions (41)	Non hypervariable regions (83)
$V_H III$ myelomas			
(Tie, Was, Jon, Zap	33	27	6
Tur, Gal, Nie) average range	28–45	25–32	3–13
Pom/Lay IgM anti-γ-globulins	8	3	9

Data from Capra and Kehoe 1974c

proteins Lay and Pom were selected for sequential analysis of the V regions (Capra and Kehoe, 1974c). These two proteins show extensive cross-idiotypic specificity, but each contains unique determinants (Kunkel et al., 1973). The proteins Lay and Pom differ at only eight positions in their V_H regions. The distribution of these differences is striking for only three occur within the statistically defined hypervariable regions. Comparing the differences between Lay and Pom with differences between any two of the seven other human $V_H III$ sequences (each of a human myeloma protein of unknown specificity), it can be seen that outside of hypervariable regions Lay and Pom are as different as any two other $V_H III$ sequences (Table 2). The correlation between antibody specificity and the hypervariable-region sequence is clear. Indeed, two complete hypervariable regions are common to both proteins. This study also underlines the value of idiotypy as a marker for hypervariable region sequence and, in some instances, for the combining site of an antibody.

(b) *Pooled guinea-pig antibody sequences.* Amino acid sequential analysis of the heterogeneous collection of V regions found in pooled normal immunoglobulin can yield a single amino acid sequence corresponding to the most prevalent amino acid at each position. The number of alternative amino acids detected at each position depends upon the sensitivity of the analysis. In such an analysis, the hypervariable regions can be detected because of the lack of any single prevalent amino acid at those steps in the sequence. On this basis, the hypervariable regions of guinea-pig heavy chains has been defined as 31–35, 48–59, and 99–118 (Birshtein and Cebra, 1971). Cebra et al. (1974) used groups of inbred guinea-pigs to raise and purify antibodies specific for three different haptens—dinitrophenyl (DNP), p-azobenzenearsonate (ARS), and p-azobenzenetrimethylammonium (TMA). A single major sequence was determined for the first 83 positions of the V regions of each of these three antibody pools. The three sequences correspond to each other and to the non-hypervariable sequence of normal pooled IgG, except that each of the antibodies shows a characteristic major sequence through the first two hypervariable regions as well as showing differences at positions 2, 16, and 79 (Figure 2). Those positions showing a different amino acid characteristic of the given antibody specificity are 16, 35, 50, 52, 54, and 59. The last five

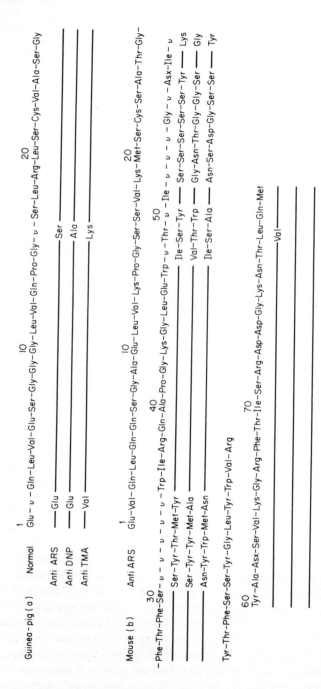

Figure 2 Sequences of anti-hapten antibodies from guinea-pig and mouse. Sequences identical to the normal protein are indicated by straight line; v indicates that no major amino acid could be detected at that position. (*a*) N-terminal amino acid sequences of V_H regions of guinea-pig normal IgGa and three antibodies to haptens. From Cebra *et al.* (1974). (*b*) N-terminal amino acid sequence of V_H region of mouse antibody to the hapten ARS; the antibody was raised in strain A/J mice and shows an idiotype characteristic of the strain. From Capra *et al.* (1975)

positions lie within the hypervariable regions statistically defined from mouse and human sequences. A correlation of hypervariable region sequence with antibody specificity can even be drawn between species. The first hypervariable region of a homogeneous mouse anti-ARS heavy chain (Capra *et al.*, 1975) also is shown in Figure 2. Of particular importance is the presence of tyrosine at position 35.

(*c*) *Mouse myeloma proteins binding phosphorylcholine.* Seven mouse myeloma proteins, each capable of binding phosphorylcholine, have been isolated independently and the V_H-region amino acid sequences determined (Figure 3). These data illustrate the subtlety of antibody diversity at the level of amino acid sequence. The overall similarity of the sequences is striking. The hypervariable regions defined for human V_H regions are marked and it can be seen that each of the proteins has the same sequence in the first hypervariable region. Most of the sequence differences between these seven proteins occur in the second and fourth hypervariable regions. Indeed, comparison of these sequences shows two short hypervariable regions from 53 to 58a and from 99 to 108. These regions show insertions and deletions as well as amino acid substitutions. The major sequence identity correlates with the similar gross specificity of these seven proteins, while the two hypervariable regions identified in proteins 603, 511, and 167 correlate with fine specificity differences between these proteins and the other four (Leon and Young, 1971). The sequences can also be correlated with the pattern of shared and unique idiotypes (see Section 4.4.7).

(*iv*) The detailed role of the hypervariable region amino acids in defining the antibody-combining site has been revealed by the determination of the three-dimensional structures of the Fab fragments of two myeloma proteins—the human protein NEW, and the mouse protein McPc 603 (Davies and Padlan, 1975; Poljak, 1975). There is a striking similarity between these two structures, the basis of which is the β-pleated structures of each domain. This structure has been termed the basic immunoglobulin fold; it also has been found to occur in two light chain dimers (Schiffer *et al.*, 1973; Epp *et al.*, 1974). In all cases, the hypervariable regions exist outside the basic immunoglobulin fold. In each of the two Fab structures, five of the seven hypervariable regions of the light and heavy chains are close to one end of the Fab, are exposed fully to solvent, and surrounding the cavity. This cavity has been defined as the antibody combining site on the basis of ligand binding. Several different ligands can be bound into the cavity of protein NEW, and subsites within the cavity have been defined (Amzel *et al.*, 1974). The contact points for each ligand are the side chains of individual amino acids within the hypervariable regions of both light and heavy chains.

This is the best evidence that the complementarity of the combining site can be determined by the pattern of amino acids in hypervariable regions. The mystery of the origins of antibody diversity will be clarified greatly if it is possible to determine the origin of hypervariable regions. A valid explanation must account

148

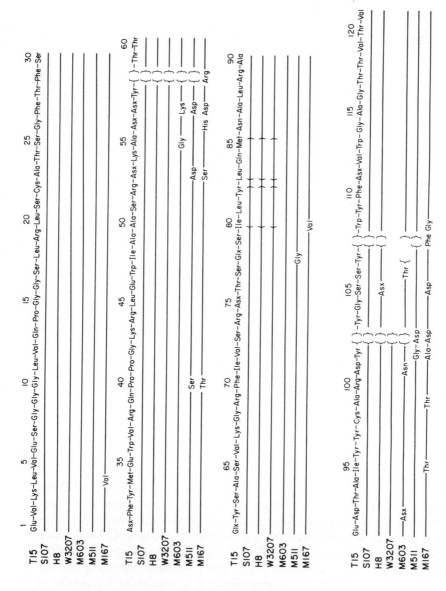

Figure 3 Complete V_H amino acid sequences from mouse myeloma proteins binding phosphorylcholine. Sequences identical to T15 are indicated by straight lines; deletions are indicated by brackets; and parentheses indicate where only compositional data is available. From Hood *et al.* (1976)

not only for the hypervariable sequences around the antibody combining site, but also for those regions remote from the combining site and playing no obvious role in defining antibody specificity.

2.3 Implications of Structure for the Origin of Antibody Diversity

At the molecular level, the origin of antibody diversity is clearly defined. The information needed to make a given antibody molecule is contained entirely within the sequences of the necessary heavy and light chains independent of any encounter with an antigen. The properties of the antibody-combining site can now be summarized, and some of the implication for antibody diversity delineated.

(*i*) The V regions of both light and heavy chains contribute to the combining site in a pseudo-symmetrical way. The arrangement of two V_L regions in a light-chain dimer also forms a pseudo-combining site (Schiffer *et al.*, 1973). The constancy (or conservative replacement) of the amino acid involved in contact between V_L and V_H regions should ensure that the maximum number of combining sites can be formed from a given number of V_L and V_H sequences (Poljak, 1975).

(*ii*) The overall size of the antibody-combining site can be varied by inserting or deleting one, or more, amino acid in those hypervariable regions proximal to the cavity. It is relevant to the origin of diversity that insertions or deletions are confined largely to those hypervariable regions surrounding the cavity.

(*iii*) In each combining site, the exact shape and specificity is determined primarily by a few (and not the same few each time) amino acid side chains rather than by all of the amino acids in the hypervariable regions. Thus, certain positions within or adjacent to each hypervariable region can be defined as complementarity-determining amino acids but these positions will differ according to the antibody. Indeed for the same V region contributing to different antibodies it is likely that different positions will contribute complementarity-determining amino acids.

(*iv*) Each antibody combining site can accommodate many structurally dissimilar compounds and can be said to exhibit multiple shared specificities. It is probable—but not proven—that for all interactions of easily measurable affinity the specific antibody combining site would contribute to a humoral immune response induced by the given haptenic group; the antibody can then be described as polyfunctional (Richards *et al.*, 1975). The prediction that identical antibody molecules will be found contributing to immune responses elicited by dissimilar haptens is consistent with available evidence. Polyfunctionality reduces the total repertoire of antibody molecules necessary to account for observed antibody diversity. This subject is discussed in greater detail by Richards *et al.* in Chapter 2 of this volume.

(*v*) A corollary to the notion of multiple shared specificities is that many different antibody combining sites can accommodate a given hapten. The nature of the interaction and the degree of fit of the hapten will vary for different combining sites, giving rise to a series of antibodies with varying affinity for a given hapten. The size of the repertoire of antibodies binding a given hapten is of particular interest for theories of the origin of antibody diversity. The prediction can be made that increasing the complexity of the hapten will place more stringent demands on the nature of the antibody combining site so that the repertoire of antibodies binding that hapten will be proportionally smaller.

3 CELLULAR BASIS OF ANTIBODY DIVERSITY

3.1 Introduction

The facts and hypotheses concerning the control of antibody synthesis at a cellular level are drawn together in this section. The details of antibody biosynthesis have been covered in an earlier chapter (see Parkhouse, Chapter 3). The historical development of the present understanding of the cellular basis of antibody diversity can be traced in the writings of Ehlich (1900), Landsteiner (1945), and Burnet (1959).

The basic question of whether information precedes antigenic encounter is answered satisfactorily at the cellular level by the clonal selection hypothesis. A selectional hypothesis operating at the molecular level was proposed by Jerne (1955), but his model still contains elements of an instructional hypothesis. The complete transition to a selectional hypothesis was made independently by Talmage (1957) and by Burnet (1957). Subsequently, the theory of clonal selection was worked out at some length by the latter (Burnet, 1959).

The clonal selection hypothesis envisages a population of cells each restricted to making a single type of antibody. Each antigen interacts with the cell or cells making a complementary antibody and the interaction of the antigen with the appropriate cell leads to cell proliferation. The resultant clonal expansion then increases production of the selected antibody. The precedent for clonal selection was the observation of selection of mutant clones in micro-organisms by appropriate environmental conditions. Thus clonal selection was linked to the idea that genes coding for antibodies are mutating and that each different clone represents a new somatic mutant. The basic hypothesis of clonal selection has survived the necessary modifications occasioned by more recent data. A current, though not necessarily final, version of the selection model is outlined here and has been reviewed in more detail elsewhere (Williamson et al., 1976).

3.2 Cells Involved in the Antibody Response

3.2.1 Stem Cells

Multipotential lymphopoietic stem cells are envisaged as a common precursor to all lymphocytes. These stem cells do not synthesize antibody and are not precommitted for the production of any particular antibody.

3.2.2 Committed Precursor Lymphocytes

These cells arise when stem cells become committed to the production of a specific antibody. The antibody is produced in small amounts (10^4–10^5 molecules/cell) and is used entirely as a receptor on the cell surface. Synthesis of receptor antibody occurs prior to, and independently of, exposure to antigen. The class of this receptor antibody is thought to be usually IgM or IgD.

3.2.3 Memory Cells

These lymphocytes are replicas of the precursor cells generated by antigen-driven clonal expansion. They differ from their progenetors mainly in being less short lived. Each is capable of giving rise to more memory cells by antigen-driven clonal expansion. Memory cells apparently can carry receptor antibody of any class or subclass.

3.2.4 Antibody-Secreting Cells

Each secretory cell produces and secretes approximately 2000 molecules of a single specificity antibody of any class or subclass per second. These cells arise by antigen-driven clonal expansion from either committed precursor lymphocytes or memory cells.

3.2.5 Conclusion

It is inherent in the above description that all cells at each stage in the expansion of a single clone produce antibody of identical specificity, although the class of antibody may vary between cells arising from the same precursor lymphocyte.

3.3 Control of Gene Expression

The essence of clonal selection is the idea of one clone—one antibody, which therefore leads to the observation of one cell—one antibody. The idea that this uniqueness of the product is due solely to mutation of antibody genes is no longer tenable (see Section 6). There are many V genes each of which can be coexpressed with any of the appropriate unique C genes (see Turner, Chapter 1 and Sogn and Kindt, Chapter 4). The restriction of antibody production in a single cell or a single clone probably involves the functional pairing of one V gene for each of the two chains of the antibody molecule. The evidence presented in the chapter on biosynthesis of antibodies (Parkhouse, Chapter 3) points to the joining of V and C genes directly at the DNA level.

The process of commitment whereby a stem cell gives rise to a precursor lymphocyte can be viewed most simply at the genetic level in terms of the joining of one V_H gene with one C_H gene, and one V_L gene with one C_L gene. The synthesis of antibody of identical specificity, but different class, by different cells arising within the same clone requires a mechanism for the movement of the selected V_H gene from its conjunction with one C_H gene to another conjunction with a new C_H gene. The simultaneous expression in one cell of the same V-gene product linked to two different C_H phenotypes requires a mechanism for duplicating the V_H gene. The mechanisms that have been suggested for the rearrangement of V and C genes have been reviewed elsewhere (Williamson and Fitzmaurice, 1976).

3.4 Diversity of Receptor Antibodies

Primotype has been defined as the total repertoire of receptor antibodies expressed on lymphocytes throughout the lifetime of an individual (Gally and Edelman, 1970). The diversity within the primotype will be much larger than the diversity of the germ-line genotype if somatic mutations contribute to the extent of antibody diversity. Whether or not the diversity of the genome in somatic cells exceeds the diversity of the genome in the germ line, it is probable that only a fraction of the primotype is expressed at any one time during the life of an individual. Most studies on the extent of the antibody phenotype utilize secreted antibodies or immunoglobulins which are the products of the ultimate cells in clonal expansion. This subfraction of antibodies available for examination thus depends upon the working of clonal selection (and in the case of myeloma proteins upon the vagaries of the neoplastic process).

The immune response to most antigens consists of a heterogeneous population of antibodies, each antibody representing the product of a single clone. Moreover, the pattern of antibodies produced in the response changes during the course of the response (Macario and Conway de Macario, 1975). In response to simple haptens, this change can be measured in terms of an increase in affinity in the later antibody. Measurements of the affinity of receptor antibodies on cells prior to exposure to the hapten suggest that receptor antibodies of both high and low affinities are present (Lefkovitz, 1974). The difference in the response between early and late periods therefore lies in the fraction of the clones which contribute antibody-secreting cells to the response. A variety of selective influences appears to direct clonal expansion either towards memory cells or towards antibody-secreting cells so that at different times during a response different clones will contribute the major portion of the secreted antibody. In certain instances, conditions may favour a single clone over all other clones. This phenomenon has been termed 'clonal dominance' (Haber, 1968; Askonas and Williamson, 1972). A special case of this is seen in the phenomenon of preferential primary selection. In this instance, the initial response to a given antigen is dominated by a particular clone (Williamson et al., 1976). Preferential primary selection is inherited as a dominant genetic trait.

4 EXTENT OF ANTIBODY DIVERSITY

4.1 Introduction

The extent of diversity must be determined at the level of the antibody phenotype and of the antibody genotype. At the level of the genome, differences between the germ-line genotype and the somatic genotype are important.

All vertebrates are capable of synthesizing antibodies. In all cases tested, antibodies could be produced against any chemical grouping. The ability of the immune response to produce antibodies against novel chemical groupings was seen initially in the work of Lansteiner (1945). These conclusions came as a surprise to those who thought of antibodies solely as a defence mechanism. Any attempt to define the extent of diversity by cataloguing the number of possible antigens is confused by the fact that each antibody also can function as a specific antigen. Therefore, a limit to the extent of antibody diversity can be set only in a closed system in which the number and diversity of antibody sites are sufficient to include an antibody specific for each of the other antibodies.

A variation in this approach of determining the extent of antibody diversity is to look for gaps in the range of immune responsiveness. Specific non-responsiveness to particular antigens can be observed but in most instances it is found not to be due to the absence of antibody of a given specificity, but rather due to the failure of clonal expansion.

In view of the many ways in which an antibody site can be constructed to fit any given epitope, and the multiple specificity of an antibody combining site, the absence of gaps in the repertoire is hardly surprising.

4.2 The Antibody Phenotype

Studies on the antibody phenotype have progressed from demonstrations of diversity, through measurements of the extent of diversity, to attempts to deduce the extent of diversity of the germ-line genotype from the measurements of the phenotype.

The diversity of the antibody phenotype has been demonstrated in terms of the enormous variety of specificities and in terms of the heterogeneity of antibodies elicited by a single antigen. The term antibody, as it is generally used, refers to a population of molecules each having in common specificity for the eliciting antigen. Individual antibody molecules having widely different affinities for the same antigen can be elicited and most antisera contain molecules having different affinities (see Steward, Chapter 7). Antibody diversity may be demonstrated by measurement of affinity constants but the number of antibody molecules of different affinity present in a population can not be deduced directly from these measurements. The extent of diversity of antibody having a defined specificity has been measured using isoelectric focusing, often coupled with biological cloning techniques, to separate and to identify each antibody species. Examples of these measurements will be given below. Idiotypy and the fine specificity of an antibody combining site are powerful markers for the identification of particular antibodies. The examples quoted below conform to the rule that as the fine specificity of the selected subset of antibodies is defined more stringently then the size of the antibody repertoire decreases. This is understandable in terms of the smaller number of ways of making an antibody combining site of a more exactly defined specificity. It is particularly important to define such small subsets of antibodies since it is then possible to determine whether the necessary V regions are encoded by germ-line genes.

The size of an antibody repertoire is estimated by measuring the frequency with which identical antibody molecules are elicited in independent events. The identity of two antibodies can only be stated with complete certainty if the complete amino acid sequences are known. Estimates of the repertoire of myeloma proteins, chosen at random independently of antibody specificity, have relied almost entirely upon amino acid sequence. For most of the studies on specific antibody repertoires, antibody identity has been assessed using as phenotypic markers isoelectric spectrotype, idiotype, or fine specificity. It is important to note that antibodies can be indistinguishable by one or more of these criteria and yet can differ in amino acid sequence.

4.3 Myeloma Protein Repertoires

4.3.1 BALB/c Mouse Myeloma Proteins

Myeloma tumours are inducible and transplantable in BALB/c mice. Considerable effort has been, and is still being, expended on sequencing the V regions of light and heavy chains of BALB/c myeloma proteins. It is assumed that using myeloma proteins obtained from a single inbred strain of mice eliminates polymorphisms which may exist within a species.

4.3.1.1 V_κ Sequences. In the mouse, κ chains constitute 95 per cent of all light chains. Partial or complete data on amino acid sequence are available for almost 50 myeloma κ chains. One way of handling these data is to organize them into subgroups each defined by sequence homology, using gaps where necessary to maximize homology. With increasing numbers of sequences being determined, the definition of subgroup has come to depend upon the genetic interpretation placed upon them (see Section 6). Hood *et al.* (1974) has pointed out that as the number of known N-terminal sequences (first 23 positions) of V_κ regions has accumulated it has not become possible to see a limit for sequence diversity being reached. However, two identical V_κ sequences have been established. The frequency of two identical sequences occurring in the 50 proteins examined allows a statistical estimate of the extent of V_κ sequence diversity; with 90 per cent confidence the number of such sequences lies between 700 and 10 000.

4.3.1.2 V_λ Sequences. Only 5 per cent of mouse light chains are of the λ-type. The V regions of 18 λ myeloma chains have been sequenced completely. This has been both possible and important because of the limited amount of diversity found in these λ sequences. Twelve of the V_λ sequences are identical, four sequences differ from this by unique single base changes, one sequence differs by two base changes, and the last differs by four base changes. The pattern of diversity is particularly interesting since all nine amino acid replacements are found in positions analogous to the hypervariable regions of the V_κ chains (Cohn *et al.*, 1974).

4.3.1.3 V_H Sequences. Only a small amount of V_H-sequence data has so far been obtained and most of it comes from myeloma proteins chosen because of their antigen-binding specificity. The tentative indication from these data is that V_H-region sequential diversity may be less than the diversity of V_κ regions (Barstad *et al.*, 1974).

4.3.2 Human Myeloma Proteins

Details of these proteins may be obtained from reviews by Hood and Talmage (1970), Pink *et al.* (1971) and Dayhoff (1972).

4.3.2.1 V_κ Sequences. In excess of 50 partial or complete V_κ sequences have been obtained. On the basis of the first 20 positions, these sequences can be divided into three subgroups (I, II, and III). A further subdivision of the V_κI subgroup also has been suggested. The sequences show considerable diversity, although this may be less than observed for the V_κ sequences in mouse. Human light chains are approximately 67 per cent κ and 33 per cent λ type.

4.3.2.2 V_λ Sequences. The 25 partial or complete V_λ sequences can be divided into four clear subgroups (I–IV). Sequence diversity is extensive and strikingly greater than in the mouse V_λ family.

4.3.2.3 V_H Sequences. More than 20 partial or complete sequences are available (Capra and Kehoe, 1974*a*). These can be classified into four subgroups on the basis of partial N-terminal sequences. Complete V-region sequences of nine V_HIII-subgroup proteins have been obtained. The concentration of attention on this subgroup is because of the unblocked N-terminal amino acid which makes the sequences immediately accessible by automated procedures. The conservation of sequential homology between proteins of the V_HIII subgroup is quite striking with two exceptions. The first exception is that variation of sequence within the hypervariable regions of this subgroup appears to be greater than that seen within the hypervariable regions of light chains. The second is that diversity increases considerably in the last 25 per cent of V_H sequences, and in fact, no subgroup specific amino acids have been found after position 82.

4.4 Specific Antibody Repertoires

4.4.1 Anti-NIP in CBA/H Mice

This repertoire was measured by using a spleen-cell transfer system at limiting cell dilution to separate the various antibody-forming clones which have been activated in mice immunized with hapten NIP (3-iodo-4-hydroxy-5-nitrophenyl) conjugated to a protein (Kreth and Williamson, 1973). A poisson distribution of memory cells was confirmed by counting the numbers of monoclonal spectrotypes of the NIP-binding antibodies present in the sera of recipient mice. Comparison of clonal spectrotypes was used to screen for similar antibodies in different donor mice. Using four donor mice, 337 spectrotypes were counted (omitting repeated occurrences of a given spectrotype from a single donor) and between donors only five pairs of spectrotype were indistinguishable. From these data the number of different NIP-binding V_H–V_L can be estimated statistically at 5000 with 90 per cent confidence limits of 2700–16 000.

This very large repertoire of antibodies showing specificity for a single hapten was rationalized by the idea of multiple-shared specificities (Williamson, 1973). This idea has gained credence with the demonstration that a single antibody combining site will bind many different antigenic determinants (Richards *et al.*, 1975). It is estimated conservatively that each antibody combining site may bind of the order of 100 different haptens. Thus, each antibody counted as having anti-NIP specificity would also be specific for 99 other epitopes. In any one mouse, something of the order of 200 different clones may be expanded by immunization with NIP–protein and 50–100 of these may contribute antibody to the response at any one time. The specificity of this population of antibody molecules depends upon the fact that they have in common a specificity for the eliciting hapten NIP. The probability that any substantial fraction of the population will share a second specificity will depend upon the degree to which the second epitope is related structurally to NIP. Thus only conventional cross-reactions predictable on a structural basis may be expected. The cross-reactivity of individual antibody combining sites for structurally dissimilar epitopes would not be observed using populations of antibody molecules.

Natural antibodies which display NIP specificity may on the basis of the foregoing reasoning, be present in normal sera in the same proportion as NIP-binding antibodies are represented in the total antibody repertoire. In a large pool of sera from non-immunized CBA/H mice, 0·09 per cent of the total immunoglobulin appears to be NIP-binding antibodies (Inman, 1974). This measurement leads to an estimate of almost 6×10^6 for the total antibody repertoire.

4.4.2 Anti-DNP/TNP in CBA/H and C3H/HE Mice

Each antibody combining site can be characterized by cataloguing the set of shared specificities for that site. If sites are selected on the basis of two shared specificities, the size of the repertoire determined should be considerably smaller than that for site selected for a single specificity. An example of this is seen in a modification of the NIP-repertoire experiment. The donor mice—either CBA/H, or the closely related strain C3H/HE—were immunized with a DNP–protein conjugate but in the recipient mice the transferred clones were challenged with TNP conjugated to the same carrier protein. The repertoire of antibody combining sites exhibiting this dual DNP/TNP fine specificity was estimated at approximately 500 (Pink and Askonas, 1974).

4.4.3 Anti-NP/NIP in C57BL/6 Mice

The fine specificity of antibody combining sites can be characterized in a more discriminating fashion by determining the binding constants of a series of structurally related compounds. This approach can be used to define a small subset of antibodies whose number then can be counted by conventional techniques. A good example of this approach is the demonstration in C57BL/6 mice of a subset of anti-nitrophenyl antibodies which show a higher affinity for NIP or NNP than for the homologous hapten (Imanishi and Mäkelä, 1974; McMichael *et al.*, 1975). Subsets with such properties have been called heteroclitic antibodies. Subsequent analysis of the spectrotypes of such antibodies shows that the repertoire of V_H–V_L combinations with this particular fine specificity is very small (certainly less than 10) with a single clone contributing predominantly to the response.

The way in which the definition of specificity of a combining site affects the observed size of the repertoire of antibodies is well illustrated by comparison of the three specific examples given above (Figure 4). On the reasonable assumption that the total repertoires for DNP in CBA/H mice and for nitrophenol in C57BL/6 mice are similar in extent to that of NIP in CBA/H mice, only 10 per cent of their repertoire can bind a second, structurally similar hapten. Only about 0·1 per cent of the repertoire of antibodies elicited by a hapten exhibits the peculiar property of a higher affinity for structurally related haptens than for the eliciting hapten. The fraction of any specific repertoire that will show binding of any other structurally dissimilar epitope is presumably similar to or less than the fraction exhibiting heteroclitic properties.

4.4.4 Anti-α-DNP–Deca-L-Lysine in Guinea-Pigs

The effectiveness of haptens, such as DNP and NIP, stems from the fact that they constitute the immunodominant portion of a complete antigenic determinant, the remainder of which is contributed by topographically adjacent parts of the carrier protein molecule. Immunodominant haptens contribute disproportionately to the binding energy when compared with the proportion of the space which they fill within the antibody combining site. This is illustrated in a

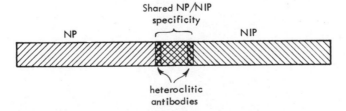

Figure 4 Probable relationships of repertoire size and overlap for antibodies to two related haptens. Heteroclitic antibodies are those elicited by one hapten but which bind the other hapten more strongly

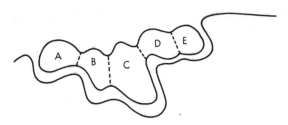

Figure 5 Model of a hypothetical antibody combining site holding a complete epitope. The parts of the epitope are labelled A, B, C, D, and E. Group C is the immunodominant part of the epitope

simplified way in Figure 5. The immunodominant hapten is designated C. Selection for the binding of C defines only one of the several subsites within the antibody combining site. If each part of the complete antigenic determinant (A, B, C, D, E) is defined exactly then the repertoire of complementary antibody combining sites should be very limited. The examples given in this and Sections 4.4.5 and 4.4.6 are consistent with this notion.

A single DNP group presented on polylysine constitutes a simple immunogen with DNP as the dominant group (Schlossman and Williamson, 1972). In guinea-pigs, α-DNP–oligolysines with a chain length of greater than seven lysines are immunogenic. The response to these oligomers, which are of the order of size expected for a complete antigenic determinant, is invariably simple consisting of one, two, or three antibodies each exhibiting a distinct spectrotype. Comparison of the spectrotypes elicited in 33 guinea-pigs of strains 2 and 13 revealed a pattern of repeats, which are catalogued in Table 3. In a total of 51 spectrotypes, only 19 different patterns could be distinguished with seven occurring several times (making up 76 per cent of the observed spectrotypes), while each of the remaining 12 spectrotypes was observed in only one response. A small repertoire of seven antibodies particularly specific for α-DNP-deca-L-lysine appears to be available

Table 3 Spectrotypes in guinea-pigs

Frequency (n)	Number* of clones (C)	$C \times n$
20	1	20
4	1	4
3	5	15
1	12	12
Total	19	51

* Number of anti-α-DNP-deca-L-lysine clones (C) found in all guinea-pigs (that is poly-L-lysine responder and poly-L-lysine non-responder animals) with a frequency n. Sera from 33 guinea-pigs immunized with α-DNP-deca-L-lysine are included in this analysis. Among the 51 monoclonal spectra compared, 19 different spectra are represented. Data from Williamson (1975)

in all guinea-pigs. The total repertoire of antibodies, which include the DNP group in their specificity, has not been determined in the guinea-pig.

Isoelectric focusing studies of antibodies elicited in response to DNP conjugated to proteins show complex spectra. In apparent contrast to this, amino acid sequential analysis of the heavy chains of such anti-DNP antibodies has failed to reveal substantial heterogeneity even in the first two hypervariable regions (Figure 2). If the total repertoire of DNP-binding antibodies is as large in the guinea-pig, as it appears to be in the mouse, then the third hypervariable region (99–118) of the heavy chain and the light chains must contribute considerably to the diversity.

4.4.5 Anti-(DNP)₂–Gramicidin S in Rabbits

Gramicidin S is a highly ordered, cyclic decapeptide made up of two identical amino acid sequences arranged in two-fold symmetry. Each half of the molecule is an ideal candidate to act as a complete antigenic determinant. For convenience, the two free amino groups have been coupled to DNP which then serves as the immunodominant part of the antigen.

This presentation of DNP is immunogenic in about 65 per cent of randomly bred rabbits (Montgomery *et al.*, 1975a, b). In each responder rabbit, the antibody obtained has a simple isoelectric spectrum. Comparison of the spectrotypes of the antibody obtained in different rabbits shows many similarities but each spectrotype is unique. Despite this, chain recombination studies using the antibodies produced in two different rabbits supports the hypothesis that the two antibodies have identical combining sites. The genetic basis of non-responsiveness remains to be proved. The intriguing possibility exists that non-responder rabbits may lack structural genes coding for antibodies capable of being elicited in response to (DNP)₂–gramicidin S. Since the evidence points to there being an extremely small repertoire of such antibodies in most rabbits, the loss of the necessary V genes in some members of the population would be probable.

4.4.6 Anti-Group A Streptococcal Polysaccharides in BALB/c Mice

Many polysaccharides contain a simple repeating pattern of monomers which, from an immunological point of view, is a multiple presentation of identical complete antigenic determinants. Antibody responses to such polysaccharides are frequently of restricted heterogeneity. Bacterial vaccines are excellent immunogenic presentations of bacterial polysaccharides.

Inbred strains of mice have been characterized as high or low-responders to each polysaccharide. Sufficiently simple responses (two to four clones) to permit analysis of the isoelectric spectrotypes have been obtained in high-responder mice by giving multiple injections of bacterial vaccine. This regime results in clonal dominance being established. Since it is not understood how the properties of a given antibody affect whether or not the clone producing that antibody can

Table 4 V-region antigenic markers (idiotypes) on phosphorylcholine-binding myeloma proteins and anti-phosphorylcholine antibodies of murine origin

Idiotype	Source of anti-idiotype serum	Ig-bearing idiotype		Location of idiotypic determinant
		Anti-PC	Myeloma	
T15	A/J or CE mice anti-T15	Only raised in Ig–1a mice (also on natural antibody Ig–1a mice)	TEPC 15 HOPC 8 M299 S63 S107 Y5170 Y5236	Not site-related (V_H site because linked to C_H -allotype)
$H8_S$	Rabbit anti-H8. Serum absorbed onto H8 and eluted with PC	Raised in any mouse	TEPC 15 HOPC 8^a CBPC 2^b	Binding-site specific, requires both V_H and V_L
H_V–PC	Rabbit anti-H8, detected by inhibition of reaction with M167	Raised in any mouse	TEPC 15 HOPC 8^a CBPC 2 M511[c] M603[c] M167[c]	Entirely on V_H, binding-site related

PC = phosphoryl choline
[a] Other myelomas in T15 group not tested
[b] CBPC2 is a PC-binding myeloma originating in a CB-20 mouse, congenic to BALB/c but with Ig–1b C_H-allotype
[c] Each of these proteins also has unique V-region determinants
Composed from Lieberman et al. (1974), Claflin and Davie (1974, 1975a, b)

achieve dominance, it must be assumed that the repertoire of dominant antibodies may be less than the total repertoire of antibodies of similar specificity. On the basis of repeated spectrotypes elicited by group A streptococcal vaccine in BALB/c mice, the repertoire of dominant antibodies was estimated to be between 30 and 40 (Cramer and Braun, 1974). In another high-responder strain, A/J mice, the dominant response showed an indistinguishable monoclonal spectrotype (A5A) in nine out of 10 mice (Eichmann, 1972, 1973). Other spectrotypes are seen in the sera of A/J mice prior to the establishment of clonal dominance.

4.4.7 Anti-Phosphorylcholine in Mice

Phosphorylcholine is a constituent of the capsular polysaccharide of certain pneumococcal strains and which acts as an immunodominant hapten. The availability of seven murine myeloma proteins showing specificity for phosphorylcholine has facilitated greatly the study of the immune response to this hapten (Claflin et al., 1975). Three idiotypes detected on phosphorylcholine-binding myeloma proteins also are present on natural anti-phosphorylcholine antibody (Table 4).

The complete amino acid sequences of the V_H regions of seven of the phosphorylcholine-binding myeloma proteins (see Figure 3) suggest a basis for the idiotypy. The first hypervariable region is identical in all seven proteins and thus correlates with the common heavy-chain idiotype H_V–PC (it should be noted that the third hypervariable region, as defined for human myeloma proteins, is common to all seven sequences). The identity of hypervariable region 2 for proteins T15, S107, and H8 correlates with the H8$_S$ idiotype, but it is known that light-chain sequences also must contribute to this idiotype. The T15 idiotype behaves as a private allotypic specificity present on anti-phosphorylcholine antibodies in Ig-1a mice but not in Ig-1b mice (Lieberman et al., 1974). The myeloma protein CBPC2 (having the C_H allotype Ig-1b) has been sequenced through position 36 of both the heavy and the light chain (Claflin et al., 1975). Comparison with the corresponding sequences of T15 shows only two differences, at positions 14 (Ser → Pro) and 16 (Arg → Gly) of the heavy chain. These differences outside of a hypervariable region could well account for the non-site-related idiotype T15.

The binding site-related idiotype H8$_S$ correlateds with the fine specificity of the antibody combining site as determined by affinity and pattern of cross-reactivity for hapten analogues (Leon and Young, 1971). By the criteria of idiotypy and fine specificity, phosphorylcholine-specific antibodies of many different inbred mouse strains and of wild *Mus musculus* share an identical combining site. On this evidence, the repertoire of anti-phosphorylcholine antibodies in any mouse would appear to be one. The existence of myeloma proteins M511, M603, and M167 points to a larger repertoire. Phosphorylcholine fills only a small portion of the combining site of M603 and is therefore not a complete antigenic determinant. This may be one explanation why M603 and the related proteins are not detected in the immune response elicited by phosphorylcholine-containing bacterial vaccines. Immunization with phosphorylcholine coupled by a tripeptide spacer to a protein does elicit a larger repertoire of antibodies (Gearhart et al., 1975). The response has been cloned and expressed in splenic fragment culture *in vitro*. Between 2 and 50 per cent of clones from different individual BALB/c mice produce phosphorylcholine-binding antibodies which lacked the T 15 idiotype.

5 ANTIBODY GENOTYPE

5.1 Deductions from Phenotype

Structural genes coding for antibodies may be mapped by conventional genetics if those genes are carried in the germ line. Studies of the inheritance of allotypic markers, amino acid sequential analysis, and molecular hybridization are each consistent with the hypothesis that a single C gene codes for each C-region sequence (see Sogn and Kindt, Chapter 4). The diversity of antibody V-region sequences led to somatic hypotheses based on the proposition that only certain basic V genes are inherited and that the phenotype represents the sum of the selected mutants of these V genes (see Section 6). It is therefore necessary to ask whether each V-region sequence is encoded by a germ-line V gene or whether it is encoded by a somatic variant V gene arising in a single cell and selected by antigen to give rise to a clone of antibody-secreting cells. Inheritance studies have used a variety of V-region markers, but usually not the most definitive one of complete amino acid sequences.

5.2. Genotype Deduced from Amino Acid Sequences

It is not feasible to investigate the germ-line basis of myeloma V-region sequences by inheritance studies. Therefore, amino acid sequences have been compared with respect to the degree and type of sequence variation in order to deduce the minimum number of germ-line V genes necessary to give rise to the observed diversity.

5.2.1 V-Region Subgroups

Once it was recognized that V-region sequences within a family are too diverse to have arisen by somatic mutations of a single V gene, the sequences were divided into subgroups. The minimum number of V genes in each family was then one for each subgroup. Subgroup assignments based on partial N-terminal sequential analysis have proved to be an oversimplification. Complete V-region sequences have been analysed by more systematic application of the subgroup principle.

5.2.2 Genealogical Analysis

V-Region sequences both within and between families have been shown to be related to each other in a genealogical pattern (Smith *et al.*, 1971; Hood, 1973). This pattern or tree is a putative map of the minimum number of genetic events necessary to have evolved the observed sequences from a single ancestral sequence. The genealogical trees drawn for antibody V regions resemble those drawn for a homologous set of proteins from many different species. The pattern obtained for homologous proteins in turn resembles the classically determined phylogenetic tree relating the species from which each of the proteins was derived. A set of V-region sequences from the same inbred strain could, it is assumed, have been expressed totally within one member of that inbred strain. In this case, each V-region sequence can be considered as arising from a single V gene whose expression is limited to a particular clone of antibody-forming cells. Thus, the genetic events relating different V-region sequences could have occurred within an evolutionary time period (V-gene duplication followed by mutational events) or during the extensive cell division necessary to produce the population of clones found within a single animal (somatic mutational events).

As with subgroup analysis, the most meaningful genealogical trees are produced using complete V-region amino acid sequences. Such an analysis for the V-regions of human myeloma proteins is shown in Figure 6. The κ, λ, and ·

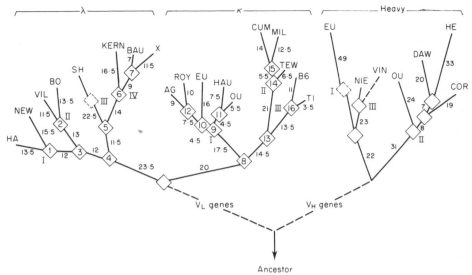

Figure 6 Genealogical tree for complete V-region sequences of the three human antibody families. The numbers of amino acid substitutions occurring between branch points are shown. Nodal sequences that have been deduced are indicated by numbers in diamonds. From Hood *et al.* (1976)

heavy-chain families can be seen to be related distantly. The division of sequences into subgroups can also be seen clearly in this analysis. In deciding upon the level at which sequence variation has arisen, genealogical analysis of myeloma sequences arising in an outbred human population are less useful than the sequences of BALB/c mouse myeloma tumours.

Genealogical analysis of the partial N-terminal sequence (23 positions) of the BALB/c κ-chain yields a tree with more than 25 branches. Only a portion of this genealogical tree can be drawn using complete V-region sequences (Figure 7). Three of these V_κ chains (M70, M321, and T124) have identical N-terminal sequences through to position 23, while M63 differs from this sequence only at position 1. Complete sequence analysis reveals that M70 differs from M321 at 22 positions, while M63 is different from M321 at only eight positions (Figure 7).

The simplest explanation of this genealogical tree is that the sequences are encoded by a minimum of three germ-line genes which arise by duplication and diversion by mutation during evolution (Hood *et al.*, 1976). This analysis is arrived at by discounting the possibility of parallel mutations, which would be required for instance if M63, M21, and T124 each represented variants of a single germ-line V gene. The argument against parallel mutations is that 'no selective forces are known which might employ random somatic mutation and select from the many variants generated two V genes with multiple identical substitutions' (Hood *et al.*, 1976). In contrast, it has been argued that structural constraints, including the antibody fold, can provide a sufficiently strong selective force for parallel mutations to be preserved (Novotny, 1973). It should however be borne

164

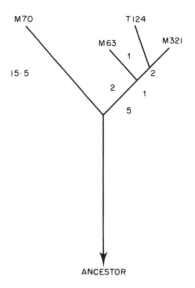

Figure 7 Genealogical tree for four mouse V regions with similar amino acid sequences. The minimum number of base changes separating each protein from the nearest common nodal ancestor is shown. From McKean *et al.* (1973)

in mind that as the requirements for a functional V-region become more stringent the wastefulness of a somatic mutation process will increase. If parallel somatic mutations are disallowed completely, most V_κ diversity must exist in the germline genome of BALB/c mice. Likewise, genealogical analysis leads to the conclusion that most of the sequence differences seen for BALB/c V_H regions require that they be coded by separate germ-line genes. For sequences as similar as the V_κ regions of M321 and T124, or the family of 18 V_λ myeloma sequences, genealogical analysis can not distinguish whether the mutational events occurred during evolution or somatically.

5.2.3 Framework and Complementarity-Determining Amino Acids

The pattern of variation seen in the V_λ sequences of 18 BALB/c mice has led to the hypothesis that the positions at which variation are seen represent complementarity-determining amino acids. This conclusion was arrived at because each of the variations occurs within one of the three hypervariable regions defined for V_κ sequences. The invariant stretches of V_κ sequence have been called the framework. The hypothesis has been advanced (Cohn, 1974; Cohn *et al.*, 1974) that 'amino acid replacements in the framework are due principally to mutation and selection of germ-line V genes, whereas replacements

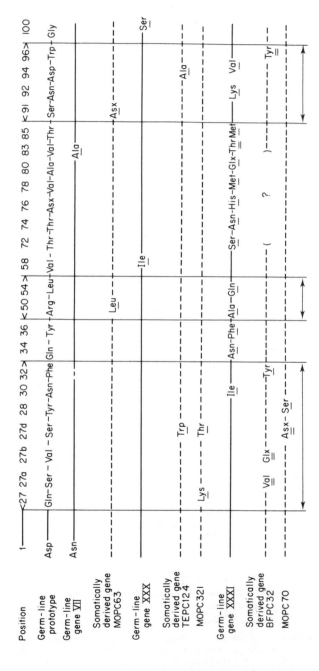

Figure 8 Predicted sequences encoded by three germ-line V$_\kappa$ genes VII, XXX, and XXXI, and the V$_\kappa$ products assumed to have been derived by mutation and selection of complementarity-determining positions (Cohn, 1974). Positions indicated ↔ are designated as complementarity-determining amino acids

of complementarity-determinants arise principally because of mutation and selection of somatically derived V genes'. In applying this hypothesis to the interpretation of other families of V-region sequences, the original argument was used to assign V_λ amino acid replacements to complementarity-determining positions has been reversed and all hypervariable region amino acids are assumed to supply complementarity determinants. The framework is taken to be those sequences separating the hypervariable regions, and a different germ-line V gene is assigned for each variation in framework sequence. From the sequences of the hypervariable regions associated with a given framework the sequence of the inherited germ-line gene is predicted. This process is illustrated in Figure 8 for the germ-line genes needed to code for the same set of proteins shown in Figure 7. The prediction of three germ-line genes is the same as the minimum estimate made on the basis of genealogical analysis.

Applying the analysis of framework variation to the known partial sequences of V_κ chains (excluding those from proteins selected previously for binding activity) a statistical estimate of the total number of germ-line V_κ genes can be made (see Table 5). Since the sequences analysed in Table 5 constitute only about a quarter of the total framework, it can be estimated that the total number of genes will be two to four times greater (150–300). Since this hypothesis was inspired by the V_λ-sequence data, it is axiomatic to the hypothesis that all 18 known V_λ sequences are the products of a single germ-line gene.

Estimates of the extent of genotype diversity based on myeloma protein sequences will be too low if myeloma proteins do not represent a random

Table 5 Distribution of unselected V_κ sequences

V_κ gene	Class	Total	Myeloma
I–XXIII	Single	23	LPC 1, MPOC 35, TEPC 157, MOPC 379, HOPC 5, MOPC 21, MPOC 63, MOPC 173, MOPC 41, MOPC 149, McPc 600, MOPC 47A, MOPC 316, BFP 61, McPc 674, McPc 843, TEPC 29, MOPC 29, TEPC 153, MPC 37, MOPC 31C, TEPC 173, 611
XXIV–XXIX	Double	12	MOPC 313, MOPC 178; MPC 11, 641 M, RPC 23; MOPC 46, MOPC 172; MOPC 467, MOPC 37; MOPC 265, MOPC 773
XXX and XXXI	Quintuplet	5	TEPC 124, MOPC 321, 613, BFPC 32, MOPC 70

$$\text{Calculated total genes} = \frac{\text{total sequences}}{2} \cdot \frac{\text{singles}}{\text{doubles}} = \frac{40}{2} \cdot \frac{23}{6} = 77$$

The sequences of only the first 23 amino acids have been compared for the mouse myeloma κ chains. A gene is assigned to each different sequence. An estimate of the total number of genes is calculated from the frequency with which sequences are found twice. This estimate of gene number must be corrected upwards to account for sequences of complete κ chains (see text). Adapted from Cohn (1974)

selection of the antibody repertoire. Comparison of available myeloma protein sequences with the pattern of amino acids found at each position in the sequence or with a normal pool of serum immunoglobulin reveals both qualitative and quantitative differences. These differences suggest that either the myeloma proteins represent a selected subset of the antibody repertoire or that a sample of 44 myeloma proteins is too few to be representative of a very extensive repertoire.

5.3 Characterization of Specific V_H Genes by Inheritance Studies

The phenotypic markers, idiotype, fine specificity, and spectrotype have been followed in breeding studies to show the inheritance of specific V_H genes. The linkage of these individual V_H genes to the C_H gene, followed by allotypic markers, is consistent with the linkage previously shown by following V_H allotypic markers in the rabbit (see Sogn and Kindt, Chapter 4). The V_H genes so far defined in the mouse are DEX(J558), T15, A5A, ARS, S117, NP(N1), and NBrP. The mapping of these genes has been described in Chapter 4 reviewed elsewhere (Eichmann, 1975).

The identification and mapping of a small number of V_H genes may account for only a tiny fraction of the germ-line V-gene pool. The distances measured by the combination between V_H and C_H genes, ranging from 0·4 to about 3 per cent, would allow space for a very large number of germ-line V genes either between the distal and proximal V genes (relative to C_H), or even between the proximal V_H gene and the C_H gene cluster.

5.4 Molecular Hybridization

Direct assessment of the nature of the germ-line V and C genes can be assessed by RNA–DNA or DNA–DNA hybridization. Several approaches are possible,

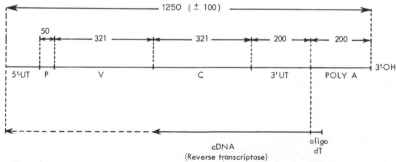

Figure 9 Sequential arrangement of the mRNA coding for MOPC 21 κ chain. The lengths of the total mRNA and its regions are given as the numbers of bases. Lengths of poly A and 3′-untranslated region (3′-UT) have been estimated from partial sequence data. The 5′UT region was estimated only by difference (Milstein *et al.*, 1974). The direction of synthesis of complementary DNA (cDNA) starting from an oligo-DT primer and catalysed by viral reverse transcriptase (RNA-directed DNA polymerase) is shown below the mRNA model

but all ideally require an isogenic mRNA preparation. Hybridization experiments using mRNA of moderate purity can lead to misleading interpretation when the nature of the impurities is unknown. The hybridization probe used may be either mRNA labelled biosynthetically or by radioiodination. Alternatively, a radioactive complementary DNA (cDNA) may be employed, which is prepared by copying mRNA using a viral reverse transcriptase (see Figure 9). There are technical difficulties in obtaining DNA transcripts complementary to the complete mRNA molecule. Partial cDNA molecules are complementary to the sequence of the 3′-untranslated region of the mRNA and the constant region of the polypeptide chain (Figure 9). Such cDNA probes have proved useful in counting the number of C genes.

Most attempts to measure the frequency of immunoglobulin genes have used the technique of hybridization driven by a vast excess of genome DNA (Melli *et al.*, 1971; Bishop, 1972). Reannealing of genome DNA follows second-order kinetics with individual gene sequences hybridizing at a rate determined by the number of copies of that gene present, thus allowing the frequency of repeated DNA sequences to be estimated. A trace quantity of a probe (mRNA or cDNA) specific for a given gene when added to reannealing DNA forms a hybrid with the complementary genomic DNA strand following pseudo-second-order kinetics and at a rate dependent on the gene frequency.

There are numerous technical problems associated with this technique. The main one is that of obtaining a probe with sufficiently high specific activity so that it can be used in the trace quantities necessary to ensure that single copy genes (representing about one part in 10^7 of genome DNA) are present in excess over the probe. When this condition is fulfilled, an approximate estimate of gene frequency can be made. For the antibody genes another problem arises, namely that the V genes in each family are homologous, but non-identical. Such mismatched sequences will cross-hybridize, although at a slower rate than the rate of annealing of identical sequences. A given V-region sequential probe can count only identical or very similar V genes, even if all V sequences are present as germ-line genes. On the other hand, if the chosen V-region probe represents a somatic variant of a germ-line gene, there will be no germ-line DNA sequence identical to the probe (Williamson and Fitzmaurice, 1976).

Hybridization experiments with probes specific for the C regions of murine κ, λ, or heavy chains reveal complementary gene sequences in non-repetitive genome DNA (reviewed by Williamson and Fitzmaurice, 1976). These results are consistent with genetic data which state that there are single copies of these genes. The data from probes including V-region sequences are less easy to interpret. The purest mRNA κ preparations used as ^{131}I-labelled probes hybridize entirely to non-repetitive DNA sequences; the extent of hybrid formation may be assessed by the resistance of the probe to ribonuclease digestion (Tonegawa, 1976). A cDNA prepared from highly purified mRNA κ (MOPC 21) and of sufficient length to include the entire V-region sequence, forms a duplex stable to deoxyribonuclease S1 with only non-repetitive DNA sequences (Rabbitts and Milstein, 1975). However, when the extent of hybridization of the same cDNA preparation was assayed by binding to hydroxyapatite, partial hybridization to repetitive DNA sequences occurred. These hybrids may represent mismatched sequences, but it is not yet clear whether the sequence involved is the V region, the 5′-untranslated mRNA sequence, or a non-κ-chain impurity in the probe preparation.

A series of experiments using $[^{131}I]$mRNA probes for the κ chains of MOPC 321 and MOPC 70 showed no hybridization to repetitive DNA—except that due to impurities in certain preparations of mRNA (Tonegawa, 1976). On the assumption that either of these probes should detect both genes (and related genes—see Figures 7 and 8), it was concluded that MOPC 321 and MOPC 70 are somatic variants of the same germ-line V gene. This conclusion is at variance with the deduction based on the analysis of amino acid sequences (see Figures 7 and 8). Calculation of the effect of sequence mismatching on the rate of hybridization of the MOPC 70 V-region sequence to the MOPC 321 V gene, or *vice versa*, suggests that separate germ-line genes coding for 70 and 321 could exist and yet not be counted by hybridization kinetics (Williamson and Fitzmaurice, 1976).

The mouse λ system also has been investigated by molecular hybridization. Leder *et al.* (1975) used a cDNA molecule of sufficient length to include most of the V_λ sequences, while Tonegawa (1976) employed $[^{131}I]$mRNA λ as the probe. In both cases, the probes hybridized to non-repetitive DNA sequences. The simplest interpretation of these results is that there is only a single V_λ gene in the mouse. The similarity of known V_λ amino acid sequences would allow a nucleic acid probe for one sequence to be used to count all gene sequences unless the gene and mRNA sequences differ by a number of silent mutations.

6 MODELS PROPOSED TO EXPLAIN ANTIBODY DIVERSITY

6.1 Introduction

In order to assess the various models proposed to explain the origin of antibody diversity, the current evidence will be summarized as a list of statements which must form the constraints of any model.

(*i*) The information required for the synthesis of an antibody molecule is present, and is expressed in the form of a receptor antibody, prior to the exposure to antigen. (*ii*) Each antibody polypeptide chain is the product of two structural genes, one coding for the V and the other for the C region. (*iii*) Each cell is programmed to make a single antibody specificity, defined by the combination of one V_L sequence with one V_H sequence. (*iv*) The joining of V and C-gene information occurs at the DNA level, *not* at the RNA or protein level. (*v*) A V gene can only be joined with a C gene on the same chromosome, that is the process is *cis*. (*vi*) There is a single C gene coding for each C_L and C_H region. (*vii*) There are three families of V genes: one linked to the cluster of C_H genes; one linked to the C_κ gene(s); and the third is linked to the C_λ gene(s). (*viii*) Each V_H gene can be expressed sequentially (and possibly simultaneously) in conjunction with each of the C_H genes.

ral of the models considered below are of historical interest only since they ~~early~~ clearly incompatible with the available evidence. Instructional hypotheses and hypotheses invoking errors in translation or transcription are not discussed, since they are no longer tenable and are irrelevant to current thinking. Each model will be discussed and its present standing evaluated.

6.2 Germ-Line Hypotheses

A germ-line hypothesis maintains that somatic processes do not play an essential part in increasing the extent of antibody diversity available from germ-line V genes. In its simplest form this hypothesis states that for each antibody polypeptide chain which an individual can synthesize a structural gene must be transmitted in the germ-line DNA (Hood and Talmage, 1970). This is held usually to be the most conventional of the hypotheses in that it assumes that each antibody gene has evolved by the same processes as the structural genes for other proteins. Successive gene duplications and subsequent diversification of the duplicates by mutation during evolutionary time are general processes in protein evolution.

The discovery of V and C-region sequences in each antibody polypeptide chain has resulted in two variations of this basic hypothesis. The first proposal was that each antibody polypeptide chain is encoded by two germ-line genes, one V and one C gene (Dreyer and Bennett, 1965). In this model, the germ-line DNA would contain a structural gene coding for each V region and one for each C region. Thus, for mouse κ chains something of the order of 1000 V-region genes and only a single C-region gene would be co-inherited. For the production of a κ chain the information from one of the V genes would need to be joined to the C-gene information. It was proposed initially that this might take place at the DNA level. In a reaction against the unconventional idea of two genes coding for a single polypeptide chain, the stringent germ-line hypothesis was restated by Brown (1972) in terms of equal numbers of V and C genes being inherited as single units. The biological precedent for the inheritance of multiple genes with alternate V and C-region nucleotide sequences came from the tandem array of 5 S RNA genes in *Xenopus laevis*. The molecular studies which have shown that there is a single haploid copy of each C gene have eliminated the stringent germ-line hypothesis (see Section 5.4).

The concept of two genes contributing to each antibody polypeptide chain now is accepted as the basis for any viable model of antibody diversity. It is not in itself sufficient for the germ-line hypothesis. On the basis of the current evidence any viable hypothesis must start from the proposition that there are multiple V genes inherited in the germ line. In the germ-line hypothesis, the term 'multiple V genes' is equated with the total repertoire. An acceptable germ-line hypothesis must, therefore, account for the pattern of diversity in V genes, the extent of the repertoire of V genes in which gaps are infrequent, and the stable inheritance of a large multiple gene family. Discussion of these points (see Section 6.4.3) will be delayed until the various somatic hypotheses have been considered.

6.3 Somatic Hypotheses Using a Small Number of Genes

The basic assumption in this class of hypotheses is that the majority of antibody diversity arises by somatic variation of a limited number of germ-line V genes. The germ-line contribution to diversity is very limited.

6.3.1 Hypermutation

A small number of V genes inherited in the germ line is subjected to special mutational events during the proliferation of stem cells. Each mutation is tested as a receptor antibody; successful mutants are those that can be selected by antigen and driven to proliferate into a clone of antibody-forming cells. The original idea was that clones of antibody-forming cells arose in a similar way to mutant clones of micro-organisms (Lederberg, 1959). As understanding of the structural basis of antibody diversity has increased so hypermutation theories have evolved to account for these data. One solution is to postulate directed mutation. Hypervariable regions might be created by an accumulation of errors in a DNA-repair process or by the random generation of short structures of DNA sequence (see, for example, Brenner and Milstein, 1966).

6.3.2 Special Selection

The alternative to directed mutation was to invoke special selection. Hypotheses of this type depend upon a special selective pressure inherent in the developing antibody system. The operation of this selection during ontogeny ensures that the necessary antibody diversity is generated by a normal mutation rate.

The most cogently argued hypothesis of this type is that of Jerne (1971), who suggested that the germ-line V genes code only for antibodies specific for the histocompatibility antigens of the species. The products of these genes would be of two types: those specific for histocompatibility antigens possessed by the individual animal; and those specific for histocompatibility antigens of the species not found in the individual animal. The special selective pressure is exhibited by self-histocompatibility antigens acting during ontogeny to suppress the expression of antibodies against such self-antigens. It was postulated that this pressure only allows the expression of mutants of the germ-line V genes originally coding for self-histocompatibility antigens. The functional immune system would comprise this selected set of spontaneous random somatic mutants together with the germ-line encoded antibodies against alloantigens.

Bodmer (1972) has pointed out a major difficulty associated with Jerne's hypothesis. The model requires a strict parallel evolution, at the population level, of genes coding for the histocompatibility antigens and for immunoglobulin V genes, but these two sets of genes are unlinked. Bodmer, therefore, modified the hypothesis of Jerne by suggesting that the V genes coding for antibodies against self-antigens are directed against differentiation antigens. These antigens characterize different differentiated cell types that are common to all individuals of a species.

simplest form of this hypothesis was the one-and-a-half gene model proposed by Smithies (1967). A single V–C pair (one gene) and a single V gene (half gene) have been proposed as the germ-line genes. Diversity of the V–C product was postulated to arise by recombination events between the two V-gene sequences. This hypothesis may be easily disproven since it predicts that only two alternative amino acids would be found at each position (except when recombination occurs within a codon).

The hyper-recombinational hypothesis of Gally and Edelman (1970) proposed that the germ-line DNA contains a tandem array of V genes which can undergo haploid, generalized recombination. This recombination would be effected simply by the formation of a V-region episome in any register, without the need for a specific excision sequence. The V-gene episome was postulated to be integrated subsequently with the nearby C gene on the same chromosome; integration was seen as being a highly specific event analogous to the integration of λ phage. Generalized recombination between a small number of V genes cannot explain the known amino acid sequences. It cannot be ruled out that recombination processes play some role in the generation of antibody diversity, but a large number of V genes must be involved. Moreover, it is necessary to divide these V genes into subgroup clusters with intra-subgroup recombination being permitted, while inter-subgroup recombination is forbidden. Even with these conditions there is no direct evidence for somatic recombination mechanisms contributing to antibody diversity.

Specific arguments against the special selection and the recombination somatic hypotheses have been discussed by Cohn (1971).

6.3.4 Clonal Variation

The accepted idea that diversity at the cellular level precedes exposure to antigen (see Section 3.4) appears to have been challenged by Cunningham (1974). His hypothesis of clonal variation consists of mutation and antigenic selection postulated to occur during antigen-driven clonal expansion. This short time scale demands a very high rate of mutation and it has been claimed that, in single clones of antibody-forming cells, variants have been observed to arise at the rate of one in 30 cell divisions. Cunningham (1974) has proposed that the variants are due to point mutations, however there is no physicochemical evidence in support of this. Two types of evidence have been offered. (*i*) Changes in antibody production by single cells have been deduced from changes in the morphology of plaques (due to lysis of erythrocytes) formed around antibody-secreting cells. The factors governing plaque morphology are complex and differences can be seen in the absence of any variation in the antibody secreted. (*ii*) Diversity of antibody-forming cells has been observed within selected clones. Such experiments do not eliminate completely the presence in the selected population of silent precursor cells that can give rise subsequently to antibody-forming cells independently of the selected clones.

None of the hypotheses seeking to explain diversity by somatic variation of less than 10 V genes is now tenable, since available sequential data require more than 100 germ-line V genes.

6.4 Current Hypotheses Using Multiple V Genes

6.4.1 Subgroups

The minimum number of V genes needed to account reasonably for known V-region amino acid sequences has increased steadily as sequential data have accumulated, and it does not appear that a limit has been reached yet. By ordering mouse V_κ sequences into a genealogical tree as many new branches for the most recently determined set of 20 sequences have been revealed as for the first 20 sequences obtained (Hood et al., 1974).

A major stage in the ordering of V-region amino acid sequential data was the recognition of subgroups of sequences sharing a similar pattern of amino acids, or gaps introduced to maximize homology. This convenient classification has been interpreted in terms of one V gene for each subgroup with intra-subgroup variation being attributed to somatic diversification processes. The definition of the degree of sequence variation necessary to define a subgroup has varied. Whatever the criterion, a V region usually has been assigned to a subgroup on the basis of only the N-terminal amino acid sequence. On this basis, the human heavy chains each fall into one of four distinct subgroups each characterized by a highly conserved N-terminal sequence. Complete V_H-region sequences show that (i) subgroup-specific amino acids are confined to the 82 N-terminal positions, and (ii) sequential diversity increases in the last 25 per cent of V_H regions (Capra and Kehoe, 1974a). The most precise definition of subgroups—and therefore the one with most predictive value—states that identical amino acid sequences outside the hypervariable regions identify sequences of the same subgroup (see Section 5.2.3). The origin of antibody diversity may then be resolved into the mechanism for generating hypervariable regions. Any valid hypothesis must account for the origin of hypervariable regions of germ-line V genes and also for subsequent localization of somatic variation, if such is postulated. Three hypotheses will be considered.

6.4.2 Somatic Mutation with Antigenic Selection

The following postulates have been advanced by Cohn et al. (1974).

(i) The germ-line complement of V genes, consisting of the order of 100 V_L and 100 V_H genes, has been selected during evolution to code for antibodies specific for pathogens. Such antibodies may afford the immediate survival of the neonate.

(ii) Selection would only act upon about 200 of the maximum number of combining sites (10^4) which the germ-line V genes could generate. In effect, all 10^4 antibodies would show germ-line inheritance although only a small fraction would be selected positively.

ιe mature animal, mutations of V genes would be expressed if a viable
nbination could be made.

ιtants of complementarity-determining amino acids will affect the
speci... ιy of the combining site and can be positively or negatively selected by
antigenic encounter.

(v) Selection must be stepwise with sequential selection acting upon each
mutation.

(vi) Ten new V genes must be generated somatically and selected to give a
minimal mature repertoire of 10^6 antibodies. A rate of 10^{-3}–10^{-5} functional
variants/cell division would be necessary (Cohn, 1974). At this rate of generation
of new V genes, it follows that in the repertoire of antibodies studied in the mature
individual the variation seen in complementarity-determining amino acids will be
largely due to somatic mutations of germ-line sequences. Hypervariable regions

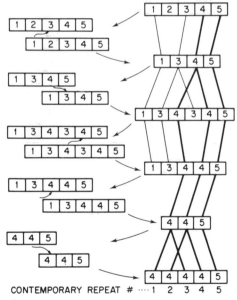

CONTEMPORARY REPEAT # ···· 1 2 3 4 5

Figure 10 Mechanism by which unequal
crossing-over can lead to the horizontal spread
of mutation (crossover fixation) and the maintenance of multiple gene sets. Each repeated
gene is represented by a box. The numbers
inside each box allow each gene to be followed
and its descendants traced. The descent of
contemporary repeats is followed by bold
lines. The original genes may be identical. A
point mutation in gene number 4 will be spread
to four out of five of the contemporary repeat
genes by unequal crossing over, thus maintaining a nearly identical set of genes (from Smith,
1973)

are assumed, therefore, to be stretches of amino acids residues whose side chains determine, or potentially determine, complementarity; in any antibody molecule, only a few side chains are involved in contact with the epitope (Davies and Padlan, 1975).

In considering this hypothesis it is important to note that direct evidence for selected somatic point mutations is lacking. Mouse λ-chain sequences are consistent with somatic mutations of a single germ-line V_λ gene, and the RNA–DNA hybridization evidence tends to support this (see Section 5.4). There is no evidence, however, that the mutant phenotypes seen have been expanded due to antigenic selection. The known variants of murine V_λ show amino acid differences at positions analogous to the hypervariable regions of mouse κ chains. However, 40 per cent of the base changes fixed in V_λ correspond to the second hypervariable region of V_κ—the region which is not involved in complementarity in either MCPc 603 or NEW combining sites.

Somatic mutations will occur at a normal rate and the accumulation of mutations in V genes will be more rapid because of the multiplicity of V genes.

6.4.3 Hypothesis for the Inheritance of a Complete Antibody Repertoire

A germ-line hypothesis for antibody diversity currently accounts for the data available (Hood et al., 1974; Williamson, 1976).

(i) Almost all V regions are encoded by germ-line genes. Some somatic variants might be expected, but these are irrelevant to the extent of antibody diversity.

(ii) The differences between V genes are the result of point mutations, deletions, and insertions accumulated during evolution.

(iii) One selective pressure for divergence of V-gene sequence is to stabilize the multiple gene family. This is achieved by decreasing the frequency with which haploid recombination eliminates V genes.

(iv) Framework (non-hypervariable) sequences are conserved by the selective pressure for V_L and V_H association.

(v) Hypervariable sequences result from the lack of selective pressure on these parts of the V-region structure. Mutations in hypervariable regions are thus neutral and are fixed at the rate at which they occur (Kimura, 1968).

(vi) One of the major forces maintaining the multiple V-gene families is unequal crossing-over between sister chromatids (Smith, 1973). The horizontal spread of mutation among a V-gene family can be accomplished by unequal crossing-over and subgroups of related V genes can be produced (Figure 10). Mutations fixed at a rate greater than the rate of unequal crossing-over will accumulate, that is neutral or positively selected mutations.

The maintenance of a multiple V gene system is currently best explained by unequal crossing-over. The idea that hypervariable regions arise due to lack of selection can apply equally to germ-line or somatic mutations. The family of murine V_λ regions could all be neutral variants of a basic sequence, irrespective of when the mutations occur, rather than being antigenically selected variants.

Maximal extimates for the number of germ-line V genes are sufficient to

r antibody diversity. The evidence for the presence of latent V genes
Kindt, Chapter 4) suggests that the number of V genes in the germ line
eed estimates based on phenotype.

6.4.4 Hypotheses Invoking Multiple Gene Interactions

In its clearest form, this type of model invokes insertion of episomes, coding for binding-site sequences, into the hypervariable regions of V genes coding for framework sequences (Wu and Kabat, 1970). The episomes might be scrambled somatically to give extensive antibody diversity.

Capra and Kindt (1974) have proposed that a set of hypervariable episomes is inherited together with individual framework V genes. This hypothesis is invoked to explain the finding of identical or cross-reactive idiotypes associated with V_H regions of different allotype in rabbits (Kindt *et al.*, 1974). The alternate explanation of these data is parallel evolution. Whatever the V-region allotype of a rabbit, a full range of antibody diversity is available. In such an extensive generation of diversity, parallel evolution of similar (see Section 2.2), or even identical hypervariable regions, is understandable and is a simpler explanation than the interaction of five or six genes to give a single polypeptide chain. Cohn (1974) has pointed out that mouse V_κ sequential data support the idea of co-evolution of hypervariable region and framework sequences, and he strongly argues against multiple gene-interaction models. Co-evolution of framework and combining site has been shown by following the inheritance of idiotype and spectrotype for the N1 V_H gene (Williamson and McMichael, 1975).

7 EVOLUTIONARY ORIGIN OF ANTIBODY DIVERSITY

7.1 Evidence from Phylogeny

The importance of antibodies can be argued from their ubiquitous presence in modern vertebrates, the most primitive representative being the cyclostomes—hagfish and lamprey—(see review by Marchalonis, 1975). Diversity of specificity is an essential feature of antibodies and is observed in all vertebrates. It is valid to examine the primitive vertebrates for gaps in the antibody repertoire, but so far there is no evidence that the range of antibody diversity is any less in the latter.

Variation in the number of V genes expressed in the λ or κ families is seen between species (Hood *et al.*, 1967). Furthermore, the proportion of V_H regions of a given subgroup is also characteristic of a species (Capra *et al.*, 1973).

Immunologically related mechanisms in invertebrates are currently under study. It is proposed to attempt to deduce the origin of the antibody system. At present, this is still a highly speculative field.

7.2 Structural Evidence

Sequence homology units (domains) appear to be the basis of evolution of all

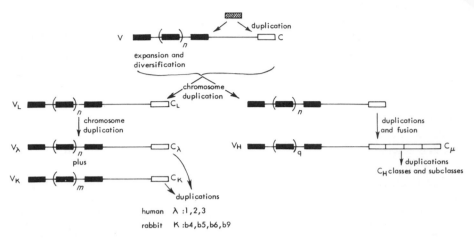

Figure 11 Putative steps in the evolution of antibody gene families from a single ancestral gene of domain size

immunoglobulin genes (see review by Hood, 1973). Possible common evolutionary pathways for modern V and C-gene families are shown in Figure 11. Gene duplication, chromosome duplication (or chromosomal translocation), and fusion of contiguous gene duplicates must be invoked. Expansion and contraction of V-gene families can be effected by unequal crossing-over (see Section 6.4.3(*vi*)), and the species differences in size of V_κ or V_λ pools can be accounted for by expansion and contraction. The same processes can spread

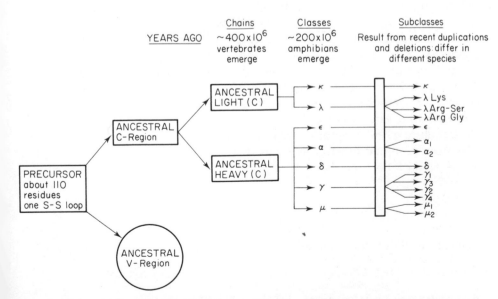

Figure 12 Scheme for the evolution of immunoglobulin C-region genes with an approximate time scale. From Milstein and Munro (1973)

rizontally, thus accounting for framework sequence identities such
t V-region allotypes (Figure 10).

cation of V genes was probably an early event in the evolution of the
ystem. Amino acid sequences of V genes from various species suggest
that divergence of V subgroups preceded speciation (Wasserman *et al.*, 1974).
Moreover, this comparison reveals that V-region sequences show greater
conservation between species than do C-region sequences. Since the V regions
contain the active site of antibodies the greater conservation of V-region
sequence parallels the conservation of active-site sequences in enzymes.

The evolution of classes and subclasses of C regions has occurred subsequent
to the divergence of V and C genes. An approximate time scale is given in Figure
12. Only IgM-type antibody is found in the most primitive vertebrates
(Marchalonis, 1975). In the chicken, ontogeny recapitulates phylogeny and IgM
production commences before IgG production, with IgA production being even
later (Cooper *et al.*, 1972).

7.3 Is There a Second Diversity System for Antigen Receptors on T Lymphocytes

The origin of antibody diversity is an important event in evolution. Nevertheless, it seems unlikely that two diversity systems could have arisen independently.
The question of a second system only arises because of the controversy over the
presence or absence of antibody-like molecules on the surface of T lymphocytes
(see Parkhouse, Chapter 3). Moreover the detection of antigen-binding molecules completely distinct from antibodies has been reported in studies on T-
helper cells (Munro and Tausig, 1975).

It is reassuring that in two systems idiotypic antigens of T-cell receptors are
indistinguishable from the idiotypes of conventional antibody of the same
specificity. Binz *et al.* (1974) have shown that an antibody raised against
alloantigen-specific T-cell receptors reacts with humoral alloantibody and blocks
T-cell reactivity in a graft-*versus*-host assay. Eichmann and Rajewsky (1975)
showed an anti-idiotypic antibody raised against A5A anti-streptococcal
polysaccharide as a carrier of hapten.

The nature of the receptor molecule carrying the idiotype remains unknown.
The presence in that receptor molecule of V gene products is strongly implied,
and it would follow that both B and T cells use the same diversity system.

REFERENCES

Amzel, L. M., Poljak, R. J., Saul, F., Varga, J. M., and Richards, F. F. (1974). *Proc. Nat. Acad. Sci., U.S.A.*, **71**, 1427.
Askonas, B. A., and Williamson, A. R. (1972). *Nature, (Lond.)*, **238**, 339.

Barstad, P., Weigert, M., Cohn, M., and Hood, L. (1974). *Proc. Nat. Acad. Sci., U.S.A.*, **71**, 4096.
Binz, H., Lindenmann, J., and Wigzell, H. (1974). *J. Exp. Med.*, **140**, 731.

Birshtein, B. K., and Cebra, J. J. (1971). *Biochemistry*, **10**, 4930.
Bishop, J. O. (1972). In *Gene Transcription in Reproductive Tissue*, p. 247.
Bodmer, W. F. (1972). *Nature, (Lond.)*, **237**, 139.
Brenner, S., and Milstein, C. (1966). *Nature, (Lond.)*, **211**, 242.
Brown, D. D. (1972). In *Molecular Genetics and Developmental Biology*, (Sussman, M., ed.), Prentice Hall, New Jersey, 101.
Burnet, F. M. (1957). *Aust. J. Sci.*, **20**, 67.
Burnet, F. M. (1959). *The Clonal Selection Theory of Acquired Immunity*, Cambridge University Press, Cambridge.

Capra, J. D., and Kehoe, J. M. (1974a). *Adv. Immunol.*, **20**, 1.
Capra, J. D., and Kehoe, J. M. (1974b). *Proc. Nat. Acad. Sci., U.S.A.*, **71**, 845.
Capra, J. D., and Kehoe, J. M. (1974c). *Proc. Nat. Acad. Sci., U.S.A.*, **71**, 4032.
Capra, J. D., and Kindt, T. J. (1974). *Immunogenetics*, **1**, 417.
Capra, J. D., Tung, A. S., and Nisonoff, A. (1975). *J. Immunol.*, **114**, 1548.
Capra, J. D., Wasserman, R. L., and Kehoe, J. M. (1973). *J. Exp. Med.*, **138**, 410.
Cebra, J. J., Koo, P. H., and Ray, A. (1974). *Science, N.Y.*, **186**, 263.
Chesebro, B., and Metzger, H. (1972). *Biochemistry*, **11**, 766.
Claflin, J. L., and Davie, J. M. (1974). *J. Exp. Med.*, **140**, 673.
Claflin, J. L., and Davie, J. M. (1975a). *J. Exp. Med.*, **141**, 1073.
Claflin, J. L., and Davie, J. M. (1975b). *J. Immunol.*, **114**, 70.
Claflin, J. L., Schroer, J. A., and Cavie, J. M. (1975). In *Proceedings of the 9th Leukocyte Culture Conference*, (Rosenthal, A. S., ed.), Academic Press, New York and London, p. 153.
Cohn, M. (1971). *Ann. N.Y. Acad. Sci.*, **190**, 529.
Cohn, M. (1974). In *Progress in Immunology II*, Vol. I, (Brent, L., and Holborow, J., eds.), North Holland Publishing Co., Amsterdam, p. 261.
Cohn, M., Blomberg, B., Geckeler, W., Raschke, W., Riblet, R., and Weigert, M. (1974). In *The Immune System: Genes, Receptors, Signals*, (Sercarz, E. E., Williamson, A. R., and Fox, C. F., eds.), Academic Press, New York and London, p. 89.
Cooper, M. D., Lawton, A. R., and Kincade, P. W. (1972). In *Contemporary Topics in Immunobiology*, Vol. I, (Hanna, M. G., ed.), Plenum Press, New York, p. 33.
Cramer, M., and Braun, D. G. (1974). *J. Exp. Med.*, **139**, 1513.
Cunningham, A. J. (1974). *Contemp. Top. Mol. Immunol.*, **3**, 1.

Davies, D. R., and Padlan, E. A. (1975). *Ann. Rev. Biochem.*, **44**, 639.
Dayhoff, M. O. (1972). *Atlas of Protein Sequence and Structure*, National Biomedical Research Foundation, Silver Spring, Maryland.
Dreyer, W. J., and Bennett, J. C. (1965). *Proc. Nat. Acad. Sci., U.S.A.*, **54**, 864.

Ehlich, P. (1900). *Proc. Roy. Soc. Secr. B.*, **66**, 424.
Eichmann, K. (1975). *Immunogenetics*, **2**, 491.
Eichmann, K., and Rajewsky, K. (1975). *Eur. J. Immunol.*, **5**, 661.
Epp, O., Colman, P., Fehlhammer, H., Bode, W., Schiffer, M., and Huber, R. (1974). *Eur. J. Biochem.*, **45**, 513.

Fisher, C. E., and Press, E. M. (1974). *Biochem. J.*, **139**, 135.
Franek, F. (1971). *Eur. J. Biochem.*, **19**, 176.

Gally, J. A., and Edelman, G. M. (1970). *Nature, (Lond.)*, **227**, 341.
Gearhart, P. J., Sigal, N. H., and Klinman, N. R. (1975). *J. Exp. Med.*, **141**, 56.
Givol, D., Strausbauch, P. H., Hurwitz, E., Wilchek, M., Haimovich, J., and Eisen, H. N., (1971). *Biochemistry*, **10**, 3461.
Goetzl, E. J., and Metzger, H. (1970). *Biochemistry*, **9**, 3862.

). *Ann. Rev. Biochem.*, **37**, 497.
). *Stadler Symp.*, **5**, 73.
stad, P., Loh, E., and Nottenburg, C. (1974). In *The Immune System: Genes, s, Signals*, (Sercarz, E. E., Williamson, A. R., and Fox, C. F., eds.). ic Press, New York and London, p. 119.
Hood, Campbell, J. H., and Elgin, S. C. R. (1976). *Ann. Rev. Genet.*, **9**, 305.
Hood, L., Gray, W., Sanders, B., and Dreyer, W. (1967). *Cold Spring Harbor Symp. Quant. Biol.*, **32**, 133.
Hood, L., and Talmage, D. (1970). *Science, N.Y.*, **168**, 325.

Imanishi, T., and Mäkelä, O. (1974). *J. Exp. Med.*, **140**, 1498.
Inman, J. K. (1974). In *The Immune System: Genes, Receptors, Signals*, (Sercarz, E. E., Williamson, A. R., and Fox, C. F., eds.), Academic Press, New York, p. 37.

Jerne, N. K. (1955). *Proc. Nat. Acad. Sci., U.S.A.*, **41**, 849.
Jerne, N. K. (1971). *Eur. J. Immunol.*, **1**, 1.

Kimura, M. (1968). *Nature, (Lond.)*, **217**, 624.
Kindt, T. J., Thunberg, A. L., Mudgett, M., and Klapper, D. G. (1974). In *The Immune System: Genes, Receptors, Signals*, (Sercarz, E. E., Williamson, A. R., and Fox, C. F., eds.), Academic Press, New York and London, p. 69.
Kreth, H. W., and Williamson, A. R. (1973). *Eur. J. Immunol.*, **3**, 141.
Kunkel, H. G., Agnello, V., Joslin, F. G., Winchester, R. J., and Capra, J. D. (1973). *J. Exp. Med.*, **137**, 331.

Landsteiner, K. (1945). *The Specificity of Serological Reactions*, Harvard University Press, Cambridge, Mass.
Leder, P., Honjo, T., Swan, D., Packman, S., Nau, M., and Norman, B. (1975). In *Molecular Approaches to Immunology*, (Smith, E. E. and Ribbons, D. S., eds.), Academic Press, New York and London.
Lederberg, J. (1959). *Science, N.Y.*, **129**, 1649.
Lefkovits, I. (1974). *Curr. Top. Microbiol. Immunol.*, **65**, 21.
Leon, M. A., and Young, N. M. (1971). *Biochemistry*, **10**, 1424.
Lieberman, R., Potter, M., Mushinski, E. B., Humphrey, W., Jr., and Rudikoff, S. (1974). *J. Exp. Med.*, **139**, 983.

Macario, A. J. L., and Conway de Macario, E. (1975). *Immunochemistry*, **12**, 249.
Mage, R., Lieberman, R., Potter, M., and Terry, W. D. (1973). In *The Antigens*, (Sela, M., ed.), Vol. 1, Academic Press, New York and London, p. 299.
Marchalonis, J. J. (1975). In *Progress in Immunology II*, Vol. II, (Brent, L., and Holborow, J., eds.), North Holland Publishing Co., Amsterdam, p. 249.
McKean, D., Potter, M., and Hood, L. (1973). *Biochemistry*, **12**, 760.
McMichael, A. J., Phillips, J. M., Williamson, A. R., Imanishi, T., and Mäkelä, O. (1975). *Immunogenetics*, **2**, 161.
Melli, M., Whitfield, C., Rao, K. V., Richardson, M., and Bishop, J. O. (1971). *Nature, New Biol.*, **231**, 8.
Milstein, C., Brownlee, G. G., Cartwright, E. M., Jarvis, J. M., and Proudfoot, N. J. (1974). *Nature, (Lond.)*, **252**, 354.
Milstein, C. and Munro, A. J. (1973). In *Defence and Recognition*, (Porter, R. R., ed.), Butterworths, London, p. 199.
Montgomery, P. C., Kahn, R. L., and Skandera, C. A. (1975a). *J. Immunol.*, **115**, 904.
Montgomery, P. C., Skandera, C. A., and Kahn, R. L. (1975b). *Nature, (Lond.)*, **256**, 138.
Munro, A. J., and Taussig, M. J. (1975). *Nature, (Lond.)*, **256**, 103.

Novotny, J. (1973). *J. Theoret. Biol.*, **41**, 171.

Pink, J. R. L., and Askonas, B. A. (1974). *Eur. J. Immunol.*, **4**, 426.
Pink, J. R. L., Wang, A.-C., and Fudenberg, H. H. (1971). *Ann. Rev. Med.*, **22**, 145.
Poljak, R. J. (1975). *Nature, (Lond.)*, **256**, 373.

Rabbitts, T. H., and Milstein, C. (1975). *Eur. J. Biochem.*, **52**, 125.
Ray, A., and Cebra, J. J. (1972). *Biochemistry*, **11**, 3647.
Richards, F. F., Konigsberg, W. H., and Rosenstein, R. W. (1975). *Science, N.Y.*, **187**, 130.

Schiffer, M., Girling, R. L., Ely, K. R., and Edmundson, A. B. (1973). *Biochemistry*, **12**, 4620.
Schlossman, S. F., and Williamson, A. R. (1972). In *Genetic Control of Immune Responsiveness*, (McDevitt, H. O., and Landy, M., eds.), Academic Press, New York and London, p. 54.
Singer, S. J., and Thorpe, N. O. (1968). *Proc. Nat. Acad. Sci., U.S.A.*, **60**, 1371.
Smith, G., Hood, L., and Fitch, W. (1971). *Ann. Rev. Biochem.*, **40**, 969.
Smith, G. P. (1973). *Cold Spring Harbor Symp. Quant. Biol.*, **38**, 507.
Smithies, O. (1967). *Cold Spring Harbor Symp. Quant. Biol.*, **32**, 161.

Talmage, D. W. (1957). *Ann. Rev. Med.*, **8**, 239.
Tonegawa, S. (1976). *Proc. Nat. Acad. Sci., U.S.A.*, **73**, 203.

Wasserman, R. L., Kehoe, J. M., and Capra, J. D. (1974). *J. Immunol.*, **113**, 954.
Williams, R. C., Kunkel, H. G., and Capra, J. D. (1968). *Science, N.Y.*, **161**, 379.
Williamson, A. R. (1973). *Biochem. J.*, **130**, 325.
Williamson, A. R. (1975). In *Isoelectric Focusing*, (Arbuthnott, J. P., and Beeley, J. A., eds.), Butterworths, London, p. 291.
Williamson, A. R. (1976). *Ann. Rev. Biochem.*, **45**, 467.
Williamson, A. R., and Fitzmaurice, L. C. (1976). In *The Generation of Antibody Diversity*, (Cunningham, A. J., ed.), Academic Press, New York and London, 183.
Williamson, A. R., and McMichael, A. J. (1975). In *Molecular Approaches to Immunology*, (Smith, E. E., and Ribbons, D. W., eds.), Academic Press, New York and London, p. 153.
Williamson, A. R., Zitron, I. M., and McMichael, A. J. (1976). *Fed. Proc.*, **35**, 2195.
Wu, T., and Kabat, E. (1970). *J. Exp. Med.*, **132**, 211.

CHAPTER 6

Immunoglobulinopathies

D. R. Stanworth

1 INTRODUCTION

Disorders grouped under the term immunoglobulinopathies have proved to be considerably more complex than was originally suspected. Initially, it was demonstrated using strip electrophoresis or ultracentrifugation that there was an overabundance or a lack of γ-globulin in the serum. With the advent of

iques, it has become customary to classify these diseases,
tative and/or quantitative abnormalities in the immuno-
main categories. This classification depends upon the
ion of levels of one or more immunoglobulin constituent.
s restricted to a single subclass or subunit, it is conveniently
ely as a monoclonal gammopathy. Emphasis in this chapter
will ~~ ~~ those immunochemical characteristics that permit such
classifications and that have been of assistance in elucidating the structure and
function of normal immunoglobulins (Turner, Chapter 1).

Another area of interest which will be considered in some depth, is the
development of laboratory models for various immunoglobulinopathies. Such
models eventually lead to a better understanding of the aetiology of the human
diseases, besides providing further valuable aids for those studying the synthesis
and role of normal immunoglobulins. These include experimental animal
systems, in which plasma-cell neoplasms and/or paraproteinaemias occur
spontaneously or are induced experimentally, and the *in vitro* culture of myeloma
cell lines that continue to secrete monoclonal immunoglobulin over long periods.

2 HYPERIMMUNOGLOBULINOPATHIES

2.1 Classical Paraproteinopathies

Although the term paraprotein was used initially (Apitz, 1940) when referring
to abnormal protein products of myeloma cells detectable in blood, urine, and
tissues, it has been applied since to any immunoglobulin-type product of an
immunocyte dyscrasia or of a malignant lymphoma. Monoclonal gammopathies, such as myeloma and Waldenström's macroglobulinaemia, are the
most prominent; and paraproteins found at high levels in the sera of such
patients have often been referred to as M proteins. This term has been extended
to include immunoglobulin subunits, such as Bence–Jones protein—light-chain
structures—produced in a relatively high proportion of patients with lympho-
plasmacytic dyscrasias, half light-chain fragments (C_1), and the products of
heavy-chain disease, comprising incomplete immunoglobulin structures detectable in the patients' serum and urine. Thus, the paraproteinopathies offer a

Table 1 Abnormal proteins in hyperimmunoglobulinopathies

Parapraproteinopathy	Abnormal protein
Multiple myeloma	IgA, IgG, IgD, IgE, Bence–Jones protein, V_L, C_L, or deleted forms
Primary Waldenström macroglobulinaemia	IgM, Bence–Jones protein
Heavy-chain (Franklin's) disease	γ, α, μ chain or deleted forms
Light-chain disease	Bence-Jones protein

rich and varied source of material for the immunochemist. These major a
malities are summarized in Table 1.

2.1.1 Incidence

2.1.1.1 Distribution among Immunoglobulin Classes and Subclasses. The frequency of documentation of paraproteinopathies depends, not unexpectedly, on the nature of the population screened and on the technique employed. The screening of blood donors, the most common group of subjects to be investigated, has revealed an overall incidence of 0·2 per cent; while a somewhat higher incidence of paraproteinopathies has been reported in other supposedly normal populations. For instance, 64 out of 6995 people (0·9 per cent) living in a Swedish county have been shown to possess an M component in their serum when it was examined by electrophoresis (Axelsson *et al.*, 1966). The distribution of this component within the three main immunoglobulin classes was: 61 per cent IgG, 27 per cent IgA, 8 per cent IgM. In serum samples from three subjects, M components of two different immunoglobulin classes were detected.

A similar ratio of IgG to IgA M component (53:25) has been observed in a relatively large group of 212 patients with myeloma (Hobbs, 1969). Only one of which was an IgD myeloma and 19 per cent were cases of light-chain disease. Of a group of 400 patients with paraproteinaemia investigated at two New York hospitals over several years, 10 per cent were proved to be suffering from Waldenström macroglobulinaemia (Osserman and Takatsuki, 1964). In a more recent study by Jancelwicz *et al.* (1975), 0·8 per cent of M components in general, were observed to be IgD; and 2 per cent of cases of myeloma in particular. The

Table 2 Comparison of features of cases of IgE myelomatosis so far documented

Patient	Sex	Age	Serum M component (g/100 ml)	Light-chain type	Bence–Jones protein	Plasma-cell leu-kaemia	Osteo-lytic lesions	Reference
N.D.	M	50	4·5	λ	+	+		Bennich and Johansson (1967)
P.S.	M	60	7·5	λ	+	+		Ogawa *et al.* (1969)
D.M.				κ				Stoice, personal communication
Hea	F	65	2·7	κ	−		+	Fishkin *et al.* (1972)
Yu	M	51		κ	−		+	Stefani and Mokeeva (1972)
Be.				κ				Penn, personal communication
K.G.				κ				Senda and Snai, personal communication
F.K.	M	59	2·1	κ	−		−	Mills *et al.* (1976)
Ka.M.	F	48	6·3	κ	−			Knedel *et al.* (1976)

the proportion of monoclonal myeloma proteins observed in
G subclass with normal serum concentrations

	IgG1 (per cent)	IgG2 (per cent)	IgG3 (per cent)	IgG4 (per cent)	Reference
	60–70		15	6	Grey and Kunkel (1964)
	77	11	9	3	Terry et al. (1965)
108	82	10	7	1	Bernier et al. (1967)
n.s.	70	18	8	3	Natvig et al. (1967)
471	78	13	6	3	Terry (1968)
147	68	14	10	8	Virella and Hobbs (1971)
121	74	10	12	4	Virella and Hobbs (1971)
848	77	14	6	3	Schur et al., unpublished results
Mean	75*	13*	8*	4*	
Normal adults	10	64	28	5	3	Leddy et al. (1970)
	n.s.	66	23	7	4	Yount et al. (1970)
		64–70	23–28	4–7	3–4	Schur (1972)
	111	72	19	8	1	Shakib et al. (1975)

* Excluding incomplete data of Grey and Kunkel
n.s. = not stated

incidence of IgE myeloma is much lower than this; only a handful of cases having been reported (Table 2). This incidence has been calculated to be of the order of 0·002 per cent of all myeloma cases on the basis of the relative levels of IgE compared with those of IgG, IgA, and IgD in sera of normal subjects.

Some assessment of the reliability of using the relative proportions of immunoglobulins in the sera of normal individuals as a guide to the incidence of different types of monoclonal gammopathy can be obtained by examining the frequency of myeloma paraproteins of different IgG subclass. Data recently obtained in the author's laboratory (Shakib et al., 1975) on the relative proportions of IgG subclasses in the sera of 111 healthy adults correspond reasonably well with the relative incidence of monoclonal IgG paraproteins of the four subclasses observed by other workers (Table 3), particularly with that reported by Natvig and associates (1967). There appear, as yet, little comparable data available on the relative incidence of myelomas belonging to different IgA subclasses.

Biclonal paraproteinaemia has been observed occasionally in cases of myelomatosis, and polyclonal paraproteinaemia has been reported even less frequently. The classification and typing of the M component in 10 multiple

Table 4 Classification and typing of M components in biclonal gammopathies

M component I		M component II		Frequency
Class	Type	Class	Type	
IgG	κ	Bence–Jones	λ	1×
IgG	κ	IgA	κ	2×
IgA	κ	IgM	κ	2×
IgG	κ	IgM	κ	1×
IgG	κ	IgG	λ	1×
IgG	λ	IgA	λ	1×
IgM	κ	IgM	λ	1×
IgG	λ	IgA	λ	1×

From Ballieux *et al.* (1968)

myeloma cases with biclonal gammopathy was studied by Ballieux and associates (1968) (see Table 4). Moreover, quantitative immunodiffusion determinations recently performed in the author's laboratory on the sera of 62 patients with IgG myelomatosis (Shakib, 1976) have revealed that 11 possessed abnormally high levels of two subclasses. The most common biclonal combination was IgG1–IgG2 (8 cases), with 2 cases of IgG1–IgG4 and 1 case of IgG2–IgG4.

2.1.1.2 Influence of Age and Sex. The incidence of paraproteinaemia increases with advancing years, reaching as high as 14 per cent in patients over 95 years old (Radl, 1974). By comparison, healthy subjects in the old-age group (60–90 years) show a frequency of around 3 per cent, but very few cases of overt myeloma. M proteins were detected in nine out of 294 Swedish subjects who were older than 70 and apparently in good health (Hällén, 1963). It has been suggested by Osserman and Kohn (1974) that, with advancing years, the number of available antibody clones may decline and therefore on stimulation only a limited number may be available, thus resulting in a restricted heterogeneity of secreted immunoglobulin.

Whereas the mean age at diagnosis of cases of IgG and IgA myeloma is in the 60s (Table 5), that of IgD (and probably IgE) myeloma seems to be somewhat younger. In a relatively large group (133) of IgD myeloma cases studied (see Table 5), 65 per cent were less than 60 years old at diagnosis; whilst the average age at diagnosis was 57 for the small group of IgE myeloma cases so far studied in depth (Table 2). IgD myeloma, unlike that involving IgG or IgA-producing plasma cells, is seen about three times more frequently in males than females. In addition, macroglobulinaemia is twice as frequent in males than in females but, in contrast to the myelomatoses, the onset of this condition occurs at an older age— between 60–80 years (Martin, 1970).

2.1.1.3 Co-existence with Other Conditions. Although the co-existence of para-proteinaemia with other pathological conditions often has been observed, the

n of clinical manifestations of IgD with other myelomas

	IgD	IgG	IgA	IgE§	Light chains only
years)	56·2	62·62*	64·65	57	56·69
	3:1	1:1	1:1	2:1	1:1
UN) value					
of 30 mg/100 ~~	67	14	33		78
Serum Ca^{2+} level of 11 mg/100 ml	30	8·33†	47–59		33–62
Mean serum M component, g/100 ml	1·7	4·3	2·8		±
Light chains of M components					
$\kappa:\lambda$ ratio	1:9	2:1	2:1	7:2	1·3:1
Bence–Jones proteinuria (per cent)	92	60	70	33	100
Osteolytic lesions (per cent)	79	57	65		78
Medium survival time,‡ (months)	9	κ 35	κ 22		κ 28
		λ 25	19		λ 11

* Two values for two different series.
† Range of means from five different series
‡ Survival time for 54 patients with fatal myeloma IgD was from the time of diagnosis; for IgG, IgA, and light chains only, the survival time was from the first nelphalan treatment
§ Summary of six well-documented cases referred to in Table 5 (modified from Jancelwicz et al., 1975)

significance of such associations is still far from clear. It is interesting that they are seen in malabsorption states sometimes; for instance, in his original description of primary macroglobulinaemia Waldenström (1944) described one patient with diarrhoea. Other macroglobulinaemic patients with this symptom have been recorded since. One such case investigated in the author's laboratory (Bradley et al., 1968) presented with steatorrhoea. Analytical ultracentrifugation of the serum (Figure 1) revealed a moderately raised serum 19 S peak. In addition, M protein, which proved to be monoclonal by light-chain analysis was detected by electrophoresis. Interestingly, ultracentrifugal analysis of the patient's serum some seven years later revealed the presence of faster-sedimenting components (29 S and 38 S) (Figure 1). This would be anticipated if the patient had Waldenström macroglobulinaemia in view of the observed increase in the level of the 19 S component to 2g/100 ml. Earlier studies (Ratcliff et al., 1963), to be referred to in more detail later (see p. 197), had shown that the faster-sedimenting components characteristic of this form of paraproteinaemia are apparent in the serum when a minimal level of the 19 S component (IgM monomer) is reached. When the patient eventually died from bronchopneumonia, necropsy revealed, in addition to the effects of malnutrition such as atrophic tongue, fatty liver, and osteoporosis, an enlargement of the spleen and extensive hyperplasia of the bone marrow. Furthermore, histological examination showed a widespread lymphocytic infiltration of all organs including the alimentary tract, resembling a reticulosis such as lymphosarcoma. However, features characteristic of Waldenström macroglobulinaemia were not apparent, despite the patient having latterly shown clinical evidence of this condition.

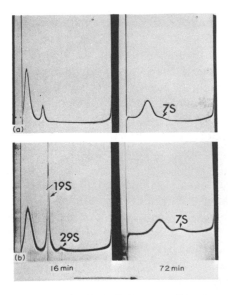

Figure 1 Analytical ultracentrifugation pattern of serum (dilution × 5): (*a*) of supposed early case of Waldenström macroglobulinaemia and (*b*) of sample of serum from same patient 6·5 years later. 59 000 rpm, 20°C, photographs taken at 16 and 72 minutes. (reproduced from Bradley *et al.* (1968)

In this connection, it is also interesting to note that a progressive rise in total serum IgA has been observed in the small number of adult patients with coeliac disease who develop a gastrointestinal lymphoma (Asquith *et al.*, 1969), but the patients apparently show no clinical evidence of myelomatosis. However, the type of pathological condition with which paraproteinaemia has been observed is by no means restricted to malabsorption syndromes; for example, 10 cases of Paget's disease with co-existing multiple myeloma have been reported recently (Srivastava and Kohn, 1974).

2.1.2 Clinical Features

The two major classes of monoclonal gammopathy, multiple myeloma and primary (Waldenström) macroglobulinaemia, are distinguishable at both clinical and cytological levels.

Clinically, cases of multiple myeloma present with bone pain, osteolytic lesions (including the characteristic 'punched-out' effect), and plasma-cell infiltration of the bone marrow. Typical clinical features of the latter disease are a tendency to haemorrhage (particularly in the gums and the retina), discrete lymphadeno-pathy, splenomegaly, and lymphocytic plasma cell infiltration of the bone marrow. In contrast to the bone picture in multiple myeloma, X-ray examina-

ltiple or focal skeletal lesions in patients with primary

ımmunocyte) level there are also distinctions between d primary macroglobulinaemia, although cytologists ut the precise origin and morphology of the neoplastic involved. According to Fudenberg (1972), many haema-

tor.... : of making a differential diagnosis between the two conditions by morphological examinations. He points out, however, that a macroglobulinaemiac patient occasionally presents with what appears to be a classical myeloma bone-marrow picture; but the marrow of such patients apparently often contains tissue mast cells, a frequent feature of macroglobulinaemia which Waldenström first drew attention to some years ago. In contrast to the plasma cells found throughout the bone marrow of myeloma patients, and which sometimes accumulate as plasmacytomas, the prominent cell type in primary macroglobulinaemia is classified less readily. In his description of macroglobulinaemia, Waldenström (1944) observed that this 'lymphocytoid' cell (sometimes referred to now as lymphocytic plasmacyte) is sessile, like the plasma cell in multiple myeloma and also occurs in the bone marrow where it rarely agglomerates into a lymphosarcoma.

Leukaemia is not seen in primary macroglobulinaemia, and has been found to be rare in multiple myeloma. For instance, of the 87 patients with myeloma seen at one group of American hospitals over a 10-year period, only three showed more than 1 per cent typical plasma cells in the peripheral smears at time of diagnosis (Ogawa and McIntyre, unpublished results, quoted by Ogawa et al., 1969). Plasma-cell leukaemia also has been seen only occasionally in the rarer form of IgD myeloma. It was of particular interest, therefore, when the first two cases of IgE myeloma to be described were reported to possess substantial plasma-cell leukaemias (see Table 2). This haematological feature, which is held generally to indicate a poor prognosis (Pruzanski et al., 1969), has not been uniformly observed in the few reported cases of IgE myeloma.

IgD myeloma shows some clinically different features to those manifest by myelomatoses involving cells producing the IgG and IgA classes, suggesting that the onset and course of the disease may be related to the structure of the M component.

Heavy-chain disease would appear to be another clinical variant of plasma-cell neoplasia. Its manifestations are primarily those of a malignant lymphoma, with little or no bone involvement. The associated abnormality in immunoglobulin metabolism involves excessive production and appearance in the serum and urine of part of a heavy chain devoid of light chains which the immunochemical aspects of this will be considered later (see Section 2.1.4.5.2).

Since Franklin and his associates (1963) described the first patient with γ heavy-chain disease, other cases have been reported involving the overproduction of α or μ, rather than γ, heavy chains. The incidence of these heavy-chain diseases is rare, with only about 100 cases having been reported in the literature since Franklin's first description. Approximately 35 of these cases were of the γ-

Table 6 Summary of seven cases of μ heavy-chain disease*

Case/Sex/ Age (years)	Chronic lymphocytic leukaemia duration (years)	Vacuolated plasma cells	Serum		Urine		Hepatosplenomegaly	Marked peripheral lympha- denopathy	Bone lesions	Amyloid
			Hypo γ	μ chain band*	Bence-Jones Protein	μ chain				
1/M/58	6	+++	+		κ		+		+	++
2/M/6	9	+++	+				+			
3/F/52	20	++	+		κ		+		?	
4/M/43	1	++	(−)		κ		+			
5/M/79	9	+	+		κ		+	+		
6/M/45	(−)	(−)	(−)	α_2		0·4 g/l	+			
7/F/48	1	?	+		κ		+			

* μ-Chain reactive protein devoid of light chains detected in all sera on immunoelectrophoresis

Reproduced from Franklin (1975)

pe, and 10 of the μ type (Franklin and Frangione, 1975).
...an et al., 1968) was first recognized in what had been
... Middle East as Arabian lymphoma of the small gut. The
...spread later to other organs and to the bone marrow. The
...features of seven reported cases of μ heavy-chain disease
...ble 6. The patients' ages ranged from 43–79 years and all
...mphocytic leukaemia. It has been pointed out by Franklin
(1975), however, that this abnormality of immunoglobulin synthesis is rare in
chronic lymphocytic leukaemia; a careful examination of over 180 patients
failing to uncover any additional cases of μ heavy-chain disease. Not unexpectedly in view of their supposed low incidence, δ and ε heavy-chain
gammopathies have not yet been reported.

Other forms of imbalance between the synthesis of heavy and light immunoglobulin chains involve the excessive production of light chains, which are excreted
as Bence–Jones protein and are usually of the same antigenic type—κ or λ—as
the intact M component. It is also interesting to note that in some cases of Bence–
Jones proteinuria, which are seen most frequently in myelomatosis (30–40 per
cent), half light-chains comprising the V_L or C_L region are found in the urine
(Cioli and Baglioni, 1966; Solomon et al., 1966). An incomplete half light-chain,
related to but clearly distinguishable from C_L has been detected in the urine of
four myeloma patients treated with corticosteroids. In other pathological
situations, referred to as light-chain disease or Bence–Jones myeloma, the
synthesis of heavy polypeptide chains appears to be blocked resulting in the
secretion only of light chain of the κ or λ type. An electrophoretically and
antigenically similar light-chain fraction is frequently detectable in the serum of
such cases.

Even in the absence of whole paraproteins, and even when these light-chain
fractions are present in extremely low levels, their presence is considered (Hobbs,
personal communication) to be an early criterion of malignancy.

2.1.3 Cellular Studies

Immunofluorescence studies of immunocytes within the bone marrow of
paraproteinopathy patients not surprisingly demonstrate a predominance of
cells secreting the monoclonal protein found in the patient's serum. In contrast,
there is a marked reduction in the number of cells synthesizing other immunoglobulin components. This accounts for the low serum levels of the other globulins
observed in the terminal stages of the disease and could, it has been suggested by
Ballieux et al. (1968), explain the high incidence of infections in the skin, and the
respiratory, and urinary tracts. Apart from studies of the frequency of
distribution of bone-marrow cells producing different types of monoclonal
immunoglobulin, attempts have been made to establish a relationship between
the morphological characteristics of the abnormal plasma cells and the actual
class of immunoglobulin being synthesized by these cells. Thus, on the basis of
cytological examination of marrow smears from 72 patients with myeloma,

macroglobulinaemia, or related paraproteinopathies, certain cell types cells' and 'thesaurocytes') were observed only in the 13 cases of IgA myelo (Paraskewas et al., 1961).

It has been suggested by Bessis et al. (1963) that the high incidence of IgA myeloma cells with large intracytoplasmic paraprotein aggregates may result from a tendency for this immunoglobulin to form high molecular weight polymers in free solution. This proposition leads to the possible clinical significance of the different polymerization patterns displayed by the monoclonal serum IgAs of different myeloma patients. This question will be considered later (see Section 4.1.4.2 and 4.1.4.3), along with the significance of other physicochemical features of serum paraproteins, such as cryoglobulinaemia and hyperviscosity manifestations.

It is interesting to note that intracytoplasmic immunoglobulin crystals also

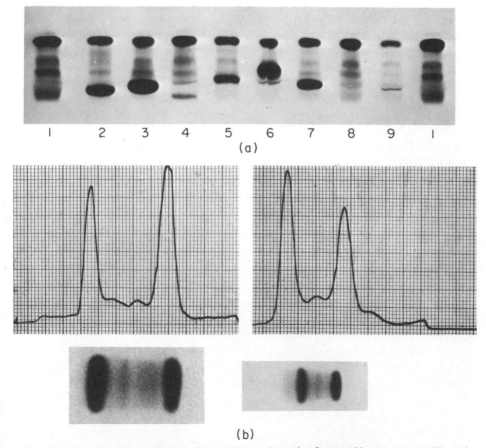

(a)

(b)

Figure 2 Comparative cellulose acetate electrophoresis of normal human serum (1), and of myeloma sera of IgG1 (2), IgG2 (3), IgG3 (4), IgG4 (5), IgA (6), IgD (8), and IgE (9), and a Waldenström macroglobulinaemic serum (7). (b) Densitometric scans of cellulose acetate electrophoretic strips of IgG1 (1) and IgG4 (2) myeloma sera

small percentage of normal and neoplastic plasma cells
...mably mirroring the capacity of some immunoglobulin
...red structures in free solution.

Aspects

2.1.4.1 *Characterization of Paraproteins.* The occurrence of a monoclonal constituent at elevated levels in the serum and/or urine is, of course, the striking immunochemical feature of classical paraproteinopathies. As is indicated by Figure 2 these are readily demonstrable in the serum by electrophoresis on cellulose acetate, although the M-component bands are usually not as pronounced in IgD or IgE myeloma sera. Immunoelectrophoretic analysis of the patient's serum often reveals a 'bowing' of the immunoglobulin precipitin line, caused by the monoclonal component, as will be seen from the examples given in Figure 3. Another frequently observed biochemical characteristic is Bence–Jones proteinuria seen in a high proportion of patients with primary macroglobulinaemia and myeloma; in particular, in those myelomas of the IgD class (Table 5), where the incidence is over 90 per cent. It is also of interest that the light chain of the M component in IgD myeloma patients is predominantly of the λ type, in contrast to the 2:1 ratio of $\kappa:\lambda$ light-chain M components seen in cases of myeloma of the other main immunoglobulin classes and in Waldenström macroglobulinaemia. In this connection, it is worth noting that it has been suggested that the clinical picture and course of multiple myeloma may be related to the class and type of M component. Certainly, it would seem of relevance that of a group of 262 cases of myeloma studied only three showed no abnormality in the serum or urine on electrophoresis or immunoelectrophoresis (Osserman and Takatsuki, 1963).

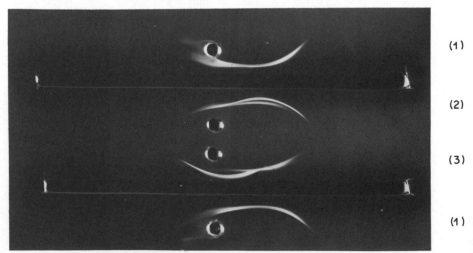

(1)

(2)

(3)

(1)

Figure 3   Immunoelectrophoretic patterns of normal human serum (1) and IgG1 (2), and IgG4 (3) myeloma sera. Antiserum: sheep anti-whole human IgG.

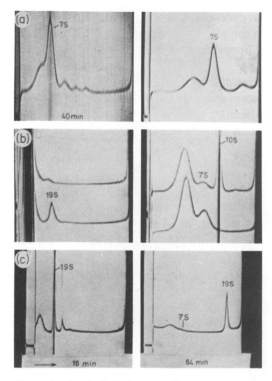

Figure 4 Analytical ultracentrifugation patterns
of (a and b) various forms of IgA myeloma sera
(dilution × 5) and (c) a typical Waldenström
macroglobulinaemic serum (dilution × 20).
59 000 rpm, 20°C. Reproduced from Stanworth
(1973a).

Ultracentrifugation is the other basic physicochemical procedure that has
thrown light on the nature of paraproteins, and this technique is important in
their classification and quantification. As far as myelomatosis is concerned, the
approach is of most value in the analysis of IgA gammopathy. In this case, the
most frequently observed pattern comprises a marked increase in IgA monomer
(7 S) with the appearance of a range of abnormal components (dimer, trimer,
etc.) sedimenting between 7 S and 19 S as is illustrated in Figure 4a. A less
frequently seen pattern involves only a gross increase in 10 S (dimer) IgA (Figure
4b), whilst a much rarer one comprises an increase in 7 S and 19 S components
without the appearance of any intermediate sedimenting forms (Figure 4c). The
significance of the polymerized forms, which are observed in the sera of certain
cases of IgA myelomatosis, and which appear to arise from inter Fc-region
disulphide bridging, is not yet understood. There is preliminary evidence
(Roberts-Thomson *et al.*, 1976) which suggests that the patients showing the
various types of polymer pattern are not readily distinguishable on clinical
grounds. Obviously, it is necessary to exclude the possibility that any additional

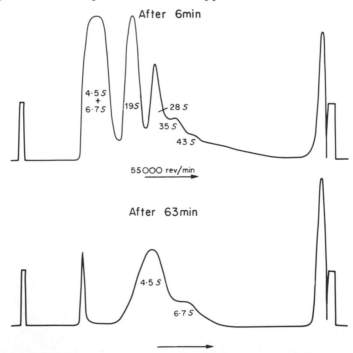

ed in the ultracentrifugation patterns of IgA myeloma ...exing with one of the many serum protein constituents ...otein, or α_1-antitrypsin) with which IgA is known to

...so has proved of some value in the identification of IgE ...ion that the sedimentation coefficient of the first such ...ich and Johansson (1967) was closer to 8 S than 7 S provided ... that it may correspond to the then unclassified reaginic antibodies found in the sera of patients with allergies of the immediate type. This suspicion was confirmed subsequently by the observation that transfer of reagin-mediated sensitivity to normal individuals' skin was blocked by prior injection of excess IgE myeloma protein (Stanworth et al., 1967). It is interesting, therefore, that myeloma IgE paraproteins isolated from cases reported more recently in Russia and Britain have proved to have lower sedimentation coefficients $(S^{\circ}_{20, \omega})$ of 7·2 S (Stefani et al., 1973) and of 7·28 S (Johns, unpublished results), respectively.

The most frequent application of ultracentrifugation has been in the identification and study of primary (Waldenström) macroglobulinaemia, in which the serum shows a characteristic ultracentrifugal pattern (Figure 4c). This comprises a greatly elevated 19 S peak which often appears as a 'concentration bar' in

Figure 5 Ultracentrifugation of a typical Waldenström macroglobulinaemia serum (diluted 1 in 5) obtained in a MSE Centriscan 75 machine

the serum Schlieren pattern. There are also faster-sedimenting compon
and 38 S) which appear to be polymerized forms of IgM covalently link
disulphide bridges probably through the C-terminal cysteine residues. In a stu
of 14 cases of Waldenström macroglobulinaemia (Ratcliff *et al.*, 1963) it was
shown that the appearance of the fast-sedimenting polymerized forms of IgM can
be expected to be seen in the patient's serum ultracentrifugation pattern when the
monomer concentration exceeds about 400 mg/100 ml for the 29 S component,
and 1·3 g/100 ml for the 38 S component (Figure 5). This has been confirmed by
repeat analysis of the patient's serum depicted in Figure 1, after an interval of
several years, by which time the concentration of the 19 S component had risen to
above the critical level for the appearance of polymer forms. Recent, more
accurate, sedimentation studies of Waldenström macroglobulinaemic sera
(Johns, unpublished results) have suggested that the $S^{o}_{20,\omega}$ values of the
detectable polymerized forms of IgM are 28, 35 and 43 S (representing dimer,
trimer, and tetramer respectively).

2.1.4.2 Hyperviscosity Effects. An advantage of ultracentrifugal analysis is that
it provides basic information about the physicochemical state of the patient's
serum to which can be related other parameters, such as viscosity, and ultimately
the clinical consequences. For instance, the increase in plasma viscosity brought
about by high levels of paraprotein can lead to a state of hyperviscosity, with
interference with essential haemostatic mechanisms. The status of the patient's
circulatory system is thought to be a decisive factor in the manifestation of the
typical haematological features of Waldenström macroglobulinaemia and

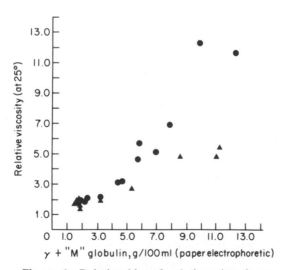

Figure 6 Relationship of relative viscosity to
paraprotein concentration of sera from patients
with myelomatosis (▲) and Waldenström macroglobulinaemia (●). Reproduced from Ratcliff *et al.*
(1963)

mmopathies. Not unexpectedly, therefore, the hyper-
ved much more frequently in those cases in which high
f monoclonal immunoglobulin are prominent—IgA
ström macroglobulinaemia. For example, a study of
bulinaemic patients revealed viscosity values in the
r cent, these high viscosities being associated usually
excess of 5 g/100 ml (McKenzie *et al.*, 1970). Likewise,
cur relatively frequently in patients with IgA myelom-
atosis, where they have been associated with high levels of a paraprotein (presumably an IgA dimer) with a high sedimentation coefficient of around 9 S (Freel *et al.*, 1972). A more recent study (Roberts-Thomson *et al.*, 1976) employed sodium dodecyl sulphate-polyacrylamide gel electrophoresis and gel filtration through Sepharose 6B to show that a high proportion (45 per cent) of 11 IgA myeloma patients, who had more than 50 per cent polymerized paraprotein in their sera, developed the hyperviscosity syndrome; in contrast to another group of 14 patients whose paraprotein was predominantly in the monomeric form, who did not display the effect.

By comparison, the incidence of the hyperviscosity syndrome is much lower in cases of IgG myelomatosis. For example, 4·2 per cent of 238 myeloma patients investigated demonstrated this feature (Pruzanski and Watt, 1972), and these represented 22 per cent of the 46 patients who had serum component levels greater than 5 g/100 ml. As will be seen from Figure 6, it is above this critical immunoglobulin level of 5 g/100 ml that the curves of serum viscosity against M component concentration for Waldenström macroglobulinaemia and IgG myeloma sera begin to part (Ratcliff *et al.*, 1963). Possibly, therefore, the

Table 7 Clinical features of myeloma observed by a number of groups investigating large numbers of cases

	IgG1	IgG2	IgG3	IgG4	Reference
Mean age at diagnosis (years)	62	60	66	66	
Bence–Jones protein	46/132	17/24	5/14	2/3	
Serum $Ca^{2+} > 10·5$ mg/ 100 ml	25/149	4/29	3/15	1/2	Schur *et al.* (1972)
BUN $\sim$ 30 mg/100 ml	26/151	4/27	3/15	0/4	
Osteolytic lesions	61/81	10/15	5/8	2/3	
Hematocrit	32	27	28	26	
Hyperviscosity	14/121	2/121	1/121	1/121	Virella and Hobbs (1971)
Hyperviscosity	5	0	4	0	Capra and Kunkel (1970) McKenzie *et al.* (1970)
Cryoglobulins	4/121	6/121	4/121	0/121	Virella and Hobbs (1971)
Survival (months)	26	28	20	16	Schur *et al.* (1972)

From Schur (1972)

hyperviscosity effects shown by certain IgG myeloma sera when the para
concentration exceeds this level can be attributed to the formation of asymm
cal IgG polymers. Indeed, two IgG myelomatosis patients with the hype
viscosity syndrome and paraprotein concentrations between 5–6 g/100 ml serum
were reported to have IgG aggregates of 11 and 14 S in their sera (Smith *et al.*,
1965). But it should be mentioned that the paraprotein was stated to be in
non-polymerized form in other cases of IgG myelomatosis with hyperviscosity
syndrome reported in the literature. Nevertheless, it is interesting that in three
myeloma patients with hyperviscosity syndrome associated with the IgG3
subclass (Capra and Kunkel, 1970) the paraprotein showed a concentration-
dependent aggregation. Furthermore, the concentration of paraprotein present
in the sera of these patients was considerably less than that in a case of
hyperviscosity associated with IgG1 paraproteinaemia, presumably because the
IgG3 molecule is the more asymmetric. Admittedly, a higher incidence of
hyperviscosity syndrome has been observed associated with IgG1 than IgG3
paraproteins (as will be seen from Table 7). But, as Schur (1972) has pointed out,
this is probably due to an extremely high serum monomer level, which incidentally
might itself be expected to favour polymerization not detected by ultracentrifugation which is of necessity carried out at lower solute concentrations.

2.1.4.3 Cryoglobulinaemia. Although cryoglobulins have been observed to
precipitate from the sera of patients with a wide range of conditions, including
rheumatoid arthritis and other connective tissue disorders, on lowering the
temperature below 37°C, they are most frequently seen on cooling the sera of
patients with multiple myeloma and Waldenström macroglobulinaemia. It is
important not to confuse cryoglobulin with cryofibrinogen which precipitates or
gels when blood is withdrawn into non-siliconized syringes containing anti-
coagulant and kept at 4° overnight. In contrast, cryoglobulin precipitates or gels
when the blood is drawn into 'dry' plastic syringes and maintained at 4° (for
preferably more than 24 hours). The resultant precipitate will normally redissolve
on restoring the temperature of the serum to 30° in contrast to isolated
cryoglobulins, which tend not to redissolve readily in physiological buffers.
Examples of the effect of temperature, pH, and ionic strength on the solubility of
cryoglobulin preparations are provided in Figure 7 which, in addition, indicates
the influence of protein concentration.

The composition of cryoglobulins, and why they tend to come out of solution
at low temperatures, has prompted considerable interest among immunochemists. In this connection, it is significant that cryoprecipitates have been
found usually to comprise immunoglobulins of different classes (IgG and IgM) or
subclasses (for example, of IgG), which has led to the suggestion that they
represent antibody–antigen like complexes. There also have been
occasional reports of cryoglobulins comprising monoclonal immunoglobulins as,
for instance, in some cases of lymphoid malignancy associated with paraproteinaemia (Stone and Metzger, 1967), and the human IgG3 protein (Cra) to
which Figure 7a refers.

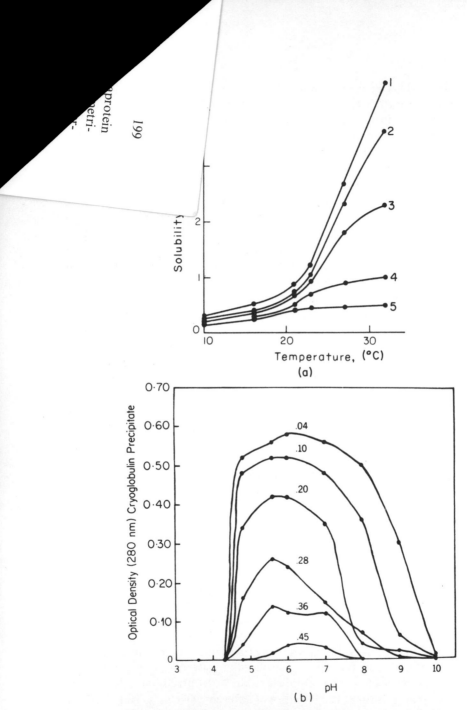

Figure 7 Effect on cryoglobulin precipitation of: (a) temperature using cryoglobulin Cra (human IgG3 myeloma protein); curves 1, 2, 3, 4 and 5 refer to cryoglobulin concentrations of: 9·40, 6·20, 3·10, 1·24, and 0·62 mg/ml, respectively; reproduced from Saluk and Clem (1975); and (b) ionic strength and pH on a cryoglobulin (human IgG myeloma protein); reproduced from Meltzer and Franklin 1966, with permission

In a preliminary study by Meltzer and Franklin (1966) of 29 patients with cryoglobulinaemia, IgG was found in the cryoprecipitate from the sera of eight cases (usually from those with multiple myeloma or idiopathic cryoglobulinaemia), IgM in nine cases, and what were described as 'unusual cryoglobulins with rheumatoid factor-like activity' in the remaining cases. Of this last group, which presented with various clinical symptoms, 11 were found to be mixed cryoglobulins containing both IgG and IgM (Meltzer *et al.*, 1966). Occasionally, cryoglobulin of the IgA class has been reported (see for example, Auscher and Guinand, 1964), but cold-precipitable globulin generally comprises IgG and/or IgM. Immunoglobulin subunits, such as Bence–Jones proteins of either light-chain type (Alper, 1966; Kiss *et al.*, 1967) and α heavy-chain disease protein (Florin-Christensen *et al.*, 1972), also have been reported to possess cryoprecipitation properties.

Important information about the mechanism of temperature-dependent gel formation has been obtained from studies on the proteolytic cleavage fragments of a single-component (6·5 S) human cryoglobulin (γ_1, λ, Gm4) from a patient with idiopathic cryoglobulinaemia (Saha *et al.*, 1970). Although papain Fc and Fab fragments of the cryoglobulin failed to form a gel at reduced temperature, the Fab fragments did exhibit a dimer–monomer equilibrium which was pH-dependent. In contrast, disulphide-bridged pepsin fragments (Fab)$_2$ formed gels in neutral and alkaline buffers at 3°. This suggests that some structural characteristic within the Fab region, probably on the Fd segment, is responsible for IgG monomer association, and ultimately for the cryogel formation. Spectro-fluorescence studies have provided evidence of localized conformational change within the Fab region during the polymerization process.

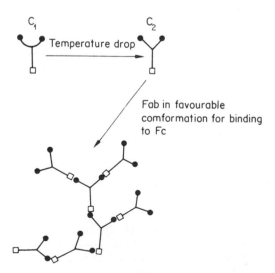

Figure 8 Postulated mechanism of cryoprecipitation of cryoglobulin Cra. Reproduced from Saluk and Clem (1975), with permission

More recent studies by Saluk and Clem (1975) on proteolytic cleavage fragments of Cra, with specificity for IgG1 and IgG3 subclasses, have prompted the proposed mechanism for cryoprecipitation (at least as far as this cryoglobulin is concerned) depicted in Figure 8. This is based on the striking correlation observed between the temperature dependence of cryoprecipitation and Fab–Fc binding, both being relatively temperature-insensitive between 10 and 21°C, but strongly temperature-dependent above 24°C. Moreover, a break in the van't Hoff binding curves at about 25°C, and the demonstration by sedimentation velocity and viscosity studies of temperature-induced conformational alteration in isolated cryoglobulin Fab fragments, suggests that a reduction in temperature below this point changes the conformational state of the Fab region to a more favourable one for the maintenance of the optimal antibody combining site conformation. When in this state, the cryoglobulin builds up large, insoluble complexes by a 'head-to-tail' combination (as is illustrated in Figure 8). The alternative possibility of Fc–Fc interaction during the original high-temperature conformational state is theoretically possible but unlikely.

It should be recognized, however, that the mechanism of cryoprecipitation may be different for cryoglobulins of different immunoglobulin compositions, even when only a single subclass is involved. For instance, in this connection, an interesting suggestion has been made by Saluk and Clem (1975), that the possession of certain allotypic (Gm) antigenic determinants by IgG molecules of the same subclass could possibly influence their functional valency of interaction with cryoglobulin IgG. In contrast to the mechanism proposed above, evidence has been obtained that the cryoprecipitation properties demonstrated by a mixed (IgM–IgG) cryoglobulin reflect the sensitivity of the solubility of the complex (between, for example, IgM antibody and effectively univalent IgG 'antigen' molecules) to changes in environmental temperature (Stone and Metzger, 1968). It has been observed that this IgM antibody, obtained from a case of Waldenström macroglobulinaemia with an atypical lymphoma, possessed functionally univalent and equivalent subunits (IgMs), and that it reacted with single determinant on IgG molecules. More recent studies on the influence of changes in temperature, ionic strength, and pH on the precipitation of a mixed Waldenström macroglobulin antibody–antigen (IgMSie–IgG) complex have indicated that the cryoprecipitation is explicable by the finite size and limited solubility of the IgM–IgG complexes (Stone and Fedak, 1974).

The mechanism whereby cryoglobulins exert their clinical significance is not understood clearly yet, although it can be supposed that in some cases immune-complex formation with concomitant complement activation plays a role in the observed vascular and tissue lesions. Meltzer and Franklin (1966) believe that the temperature at which precipitation begins is more important than the total serum cryoglobulin content. Many patients were observed to be asymptomatic with levels at 1·5–1·8 g/100 ml, while others with as little as 0·1–0·6 g/100 ml showed symptoms attributable to the cryoglobulin. It seems conceivable that the precise composition of the cryoprecipitate could be critical as far as the possibility of activating Fc-located effector sites is concerned.

Apparently when the serum cryoglobulin concentration exceeds 2 g/100 ml, it becomes difficult to separate the problems secondary to the increase in serum viscosity from those primarily due to cold precipitation. Indeed it has been suggested by Florin-Christensen (1974) that when serum levels of cryoglobulin rise above 1–2 g/100 ml in a symptomatic patient daily plasmaphoresis in a 37° circuit should be performed. An alternative form of treatment has involved a relatively long course of oral D-penicillamine, a thiol compound which has also been used occasionally in the treatment of Waldenström macroglobulinaemia to prevent the build up of high levels of circulating macroglobulins but which is used more widely in the treatment of Wilson's disease and rheumatoid arthritis. In one such report, a marked reduction in the cryoprecipitation shown by a patient with essential (IgG–IgM) cryoglobulinaemia was observed, but this was not accompanied by any clinical improvement (Goldberg and Barnett, 1970).

Obviously, further physicochemical and immunochemical studies are required on human cryoglobulins of differing composition and specificity. Moreover, valuable information about factors influencing the synthesis of cryoglobulins and the structural basis of their precipitability in the cold should be forthcoming from studying experimentally-induced cryoglobulinaemias in animals. Such a model has been established, in New Zealand Red rabbits immunized with streptococcal Group B/C vaccines (Herd, 1973a, b). Cryoglobulins begin to appear in the serum about three weeks after the commencement of immunization, and reach maximal levels ranging from less than 0·1 to greater than 6 mg/ml. The isolated cryoglobulins thus produced have been shown to comprise IgM and IgG antiglobulins (mainly of the IgM class), and homogeneous IgG antibodies against streptococcal antigens and DNA.

2.1.4.4 Paraproteins with Antibody Activity. The occurrence of circulating monoclonal immunoglobulins with antibody activity in some patients with hyperimmunoglobulinopathies is now well established. Apart from the anti-γ-globulin (rheumatoid factor-like) activity often manifested by cryoprecipitation (as discussed in Section 2.1.4.3), cold-agglutinin (anti-I and anti-i) activity is probably the most frequently observed. Sometimes even both reduced temperature effects are expressed in the same case of Waldenström macroglobulinaemia. Naturally occurring cold agglutinins, like those that develop during infections, are usually polyclonal immunoglobulins of the IgM class possessing both κ and λ light-chain types (Harboe and Lind, 1966). In contrast, the cold agglutinins found in the sera of patients with lymphoproliferative disorders are monoclonal IgM proteins, which were thought originally to be exclusively of the κ type (Harboe et al., 1965), but which have been shown more recently to be sometimes of the λ light-chain type. Feizi (1967), for example, has reported a disorder of the latter type in three patients with chronic haemolytic anaemia.

Another widely investigated aspect is the hapten-binding activity of monoclonal immunoglobulins occurring in the sera of patients with multiple myelomatosis, Waldenström macroglobulinaemia, and related disorders. Indeed, such paraproteins have been employed widely by immunochemists as model 'antibodies' in the study of kinetic and structural aspects of antigen–(hapten)-binding interactions. Apart from the academic value of this approach, it would seem to be of some practical significance. This is indicated by the interesting observation of Bauer (1974) that three Waldenström macroglobulins have been found which precipitated derivatives of triiodoaminobenzoic acid present in the contrast media used for cholecystography. Intravenous injection of a derivative resulted in a fatal immunological reaction in one Waldenström macroglobulinaemic patient, whose IgM was shown to have 'hapten'-binding activity localized within the Fab regions. Another 'antibody'-like Waldenström macroglobulin, which has been studied in some depth, possesses specificity against phosphorylcholine (Reisen, 1975).

It is imperative in the identification of paraproteins with antibody activity to demonstrate that the presumptive ligand is bound to the monoclonal immunoglobulin *via* its Fab region. This applies particularly to suspected cases of specific binding by paraprotein of lipoprotein, which has been reported in cases of myeloma with accompanying hyperlipaemia and xanthomatosis, and of other plasma protein constituents such as albumin. Some immunoglobulins (for example, IgA) have been observed to bind these plasma proteins non-specifically.

Other instances of monoclonal immunoglobulins with antibody activity include anti-tissue antibodies, such as those directed against microsomes, mitochondria, smooth muscle, and foetoproteins in acute and chronic liver disease (Florin-Christensen and Roux, 1974). Membrane-bound monoclonal immunoglobulins with anti-IgG, anti-Forsmann, and cold-agglutin activity also have been reported (Seligman and Feizi, 1974). Interestingly, the membrane-bound IgM of a variable proportion of small lymphocytes in peripheral blood of such patients has been shown to possess the same antibody specificity and idiotype as the monoclonal IgM in the serum—where detectable. This has prompted the suggestion by Harboe and Feizi (1974) that monoclonal populations of lymphocytes from patients with chronic lymphatic leukaemia may prove to be as useful in cellular immunological studies as monoclonal immunoglobulins isolated from the serum of patients with myelomatosis and Waldenström's macroglobulinaemia have been in the delineation of immunoglobulin structure.

Important information about the nature of paraproteins with antibody activity is being obtained also from studies on immunoglobulins with restricted heterogeneity induced experimentally by immunization of animals such as rabbits. This aspect will be discussed later when considering experimental models (Section 4.1.1).

2.1.4.5 Depleted Paraproteins

2.1.4.5.1 Heavy-chain disease proteins. The fore-runner of the increasingly available 'library' of spontaneously occurring depleted paraproteins was the heavy chain-like protein first reported by Franklin (1964). This protein was detected in the serum and urine of a 43-years old male patient who was admitted to hospital initially with generalized lymph node enlargement, but who showed evidence of a malignant lymphoma eventually. A large, abnormal homogeneous peak of fast γ mobility (8·4 g/100 ml) was observed on electrophoresis of the patient's serum. A similar dominant electrophoretic peak was observed, along with lesser amounts of albumin and globulin in the urine, approximately 12 g/day being excreted. Preliminary analyses showed it to possess a sedimentation coefficient ($S^{\circ}_{20,\omega}$) of 3·7 S, a molecular weight of 53 000 when determined by the Archibald method, and conjugated carbohydrate. It was found also to possess Gm-genotype specificity, but to lack light-chain InV specificity.

This initial observation was followed soon after by a report by Osserman and Takatsuki (1964) of four further cases of this so-called heavy-chain disease in the U.S.A. with many clinical and biochemical features similar to the original case. Many further examples of this condition, as well as of the production of defective α and μ heavy chains, have been recorded since. For instance, a survey of the literature reported 35 cases of γ; 50 of α, and 10 of μ heavy-chain disease (Franklin and Frangione, 1971). A subdivision of heavy-chain diseases, on the basis of immunological studies, has revealed 68 per cent to be of the γ_1 subclass, 7 per cent of the γ_2, 21 per cent of the γ_3, and only 4 per cent of the γ_4 (Frangione and Franklin, unpublished results). The reason for the unexpectedly high incidence of γ_3 heavy-chain disease proteins is not understood yet.

Table 8 Major types of abnormal heavy chains observed so far in heavy-chain diseases

	Variants	Species*	
Heavy chain	Heavy-chain disease proteins (γ, α, μ)—altered heavy chains— usually no light chains		
	Internal deletion	H	?M
	Degradation	H	
	Intact heavy chain and probable degradation	H	M
	Free heavy and light chains—unassembled	H	
	Whole molecule with altered heavy chain		
	Internal deletion like heavy-chain disease		M
	Missing hinge	H	
	Missing C_H3 domain	H	M
	Missing Fc μ fragment	H	
	Half molecules with deletion missing plus an additional change		M
	Degradation of N-terminal domain	H	
	Hybrid molecules and longer heavy chains	H	M
Light chain	Myelomas with altered light chains		
	Internal deletion	H	
	Elongated light chain	H	M
	Light-chain deletions or fragments		
	Internal deletion and carbohydrate	H	
	C-region only		M
	Amyloid—immunoglobulin related	H	

* H = human, M = murine

From Franklin and Fraugione (1973)

Subsequent immunochemical analyses have shown that while the original heavy-chain disease provides the most striking abnormality in immunoglobulin synthesis, a series of other more subtly altered molecules is produced under certain pathological conditions, or experimentally in mice. The major types of abnormality of immunoglobulin chains so far observed are listed in Table 8.

Although initial immunochemical studies suggested that heavy-chain proteins were essentially Fc fragments, detailed amino acid sequential analyses have revealed (as illustrated in Figure 9) a wide variety of structural abnormalities. Such abnormalities range from apparently intact free heavy chains in one instance to proteins with internal deletions of part, or most, of the V region as well as the C_H1 domain up to, and sometimes including, the hinge region. A number of molecules has been found to lack the whole of the variable region and the C_H1 domain (Franklin and Frangione, 1971). Moreover, the discovery of a γ_3 heavy-chain disease protein (OMM) existing in two forms, namely the intact heavy chain as well as the Fc fragment including the hinge region (Aldersberg *et al.*,

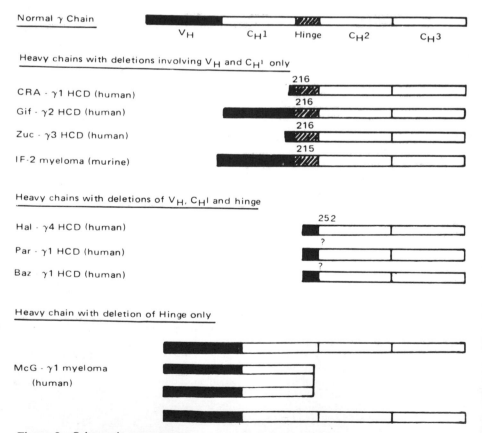

Figure 9 Schematic representation of various types of deletion seen in heavy-chain disease and myeloma proteins with internal deletions, compared to normal heavy chain. Reproduced from Franklin and Frangione (1975), with permission

unpublished results), has led to the suggestion that such structures probably result from proteolytic degradation of a larger precursor molecule.

Two major types of deleted heavy-chain disease proteins are now recognizable (see Figure 9): (*i*) those with internal deletion of part of the V region and the C_H1 domain, with resumption of normal sequences at glutamate 216; and (*ii*) those with internal deletion of the V region, the C_H1 domain, and the hinge region. In addition, other γ heavy-chain disease proteins have been found to be deleted only at the hinge region. Others begin at the hinge region and comprise the whole of C_H2 and C_H3 domains, thus appearing to be degradation products. Biosynthetic studies by, for example, Buxbaum (1973), employing tissue culture of cells from patients with heavy-chain (γ, α, or μ) disease, have failed to produce evidence of post-ribosomal degradation, thereby suggesting a synthetic origin for the abnormal proteins. Moreover, the results of the study of the immunoglobulin products of spontaneously arising structural gene mutants in a cultured cell line of the mouse myeloma MOPC 21 strain would seem to be consistent with this possibility. Such proteins display an internal heavy-chain deletion in a position (residue 215) homologous to the one (residue 216) implicated in the majority of internal deletions observed in human heavy-chain disease (Milstein *et al.*, 1974).

α and μ heavy-chain disease proteins are not yet as well characterized as the γ heavy-chain immunoglobulins, but an essentially similar picture is beginning to emerge (see Table 9). For instance, the α heavy-chain disease proteins resemble the γ proteins in the size of the monomer form (35 000–42 000), in comprising primarily the C-terminal half of the heavy polypeptide chain, and in being rich in carbohydrate. Furthermore, all proteins so far studied lack heavy-light chain disulphide bridges, while retaining part or all of the heavy–heavy disulphide bridges and possessing a normal C-terminus (Seligmann *et al.*, 1971). Interestingly, all are of the major $\alpha1$ subclass. They also seem to contain J chain, as is indicated in Table 9. A quite different IgA abnormality seen in three generations

Table 9 Properties of γ, α, and μ heavy-chain disease proteins

Property	γ	α	μ
Mobility	Fast γ-β	β	$\alpha1$
Sedimentation rate	2·8–4 S	4–11 S	10·8–11 S
Molecular weight of monomer	25 000–58 000 (most 35 000)	35 000–42 000	35 000–55 000
Carbohydrate	Rich–up to 20 per cent	Rich	Rich
Present in urine	Usually	Sometimes	Never
Light-chain production	Not yet observed	Not yet observed	5/7
J chain	No	Often	In some
Approximate number of cases	35	50	10
Subclass	$\gamma1$ 19/28 $\gamma2$ 2/28 $\gamma3$ 6/28 $\gamma4$ 1/28	All $\alpha1$	—

From Franklin and Frangione (1973)

of a family in which the propositus suffered from recurrent infections (Moroz *et al.*, 1971) would seem to represent a familial defect of IgA1 heavy and light polypeptide chain assembly.

A smaller number of μ heavy-chain disease proteins have been studied so far. It is noteworthy (see Table 6) that six out of seven were found to have chronic lymphocytic leukaemia (Franklin, 1975) and that two features common to all were the presence of κ type Bence–Jones proteins and vacuolated plasma cells in the marrow. The μ chain protein with a molecular weight of around 55 000, compared with about 6000 for the normal, is a synthetic not a degradation product (Franklin, 1975). The cells appear to synthesize an incomplete μ chain and a light chain, which may be due to a mutation giving rise to an internal deletion of the μ chain so that it cannot be joined into a complete macroglobulin molecule.

Other types of altered chain seen in myeloma proteins of human and murine origin include: deletions of the C-terminal end of the heavy chain (in a human IgA myeloma); the omission of the whole Fc region of μ (that is C_H3 and C_H4 domains); half molecules (for example of IGA) with deletion; internal deletion of light chain with loss of the V region of the heavy chain (for example in an IgG myeloma protein); and molecules which appear to be hybrids of two different subclasses, as in the case of murine IgG2a and IgG2b. An example of the last abnormality was the report by Kunkel and associates (1969) of human IgG molecules in normal individuals comprising γ_3 Fab and γ_1 Fc fragments. Reports of other, even more subtle examples of hybridized normal and myeloma human IgG molecules are beginning to appear in the literature.

2.1.4.5.2 Light-chain variants. Apart from the synthesis and secretion of free light chains in hyperimmunoglobulinopathic conditions, similar abnormalities to those outlined above have been encountered involving both structural deletions and alterations. These too are listed in Table 8.

Minor internal deletions are, of course, often revealed in comparisons of the primary amino acid sequences of different light-chain proteins by the necessity of leaving short gaps to permit proper alignment. Of greater interest, however, are reports of omission of whole V and C regions for synthetic or degradative reasons, the resultant product being sometimes distinguishable by antigenic analysis and peptide mapping from a normal light-chain domain. An example of such a protein is the newly recognized form of light chain in four cases of myelomatosis treated with intermittent corticosteroids (Solomon *et al.*, 1966). This was associated with a marked but transient decrease in the 'complete' Bence–Jones proteinuria along with the transitory appearance of the new protein (designated C*). Similar light-chain products have been shown to be produced by mouse myeloma cell lines, and should throw important light eventually on the genetic control of light-chain synthesis and the reasons for the appearance of the aberrant forms in human disease states.

Other types of light-chain variants include elongated light chains, and myeloma proteins with internally deleted light and heavy chains, such as the human IgG myeloma protein (SM) which has been found to possess deletions of similar magnitude (molecular weights of approximately 10 000) in both V regions (Isobe and Osserman, 1974). This abnormal protein with a molecular weight of around 110 000) was detectable at levels of 5–6 g/100 ml in the sera along with the free Fc fragment. The latter also was seen in the urine together with a daily urinary secretion of 10–20 g of deleted light chains with molecular weight of about 15 000.

It would seem obvious that the types of immunoglobulin variants described here will prove of increasing value as 'structural tools' besides providing further clues about the factors controlling immunoglobulin synthesis. Hopefully, their further study will throw more light on the influence of plasma cell proliferative and lymphoproliferative disorders on normal antibody production.

2.1.5 Amyloidosis

The nature of amyloid is dealt with in detail by Pras and Gafni (Chapter 14): Nevertheless, a brief account is included here for completeness.

Although amyloid deposits have been observed in the tissues in a variety of conditions, they are seen frequently in association with Bence–Jones proteinuria in patients with multiple myeloma. Initially, they were thought to consist mainly of carbohydrates (hence the term amyloid—meaning starch-like) but were found later to be of a proteinaceous nature. More recent studies (summarized by Glenner et al., 1972) provided convincing evidence that amyloid comprises, at least in part, deposits of light-chain related immunoglobulin fragments. But non-immunoglobulin components, including polysaccharide, appear to be present in many if not all 'secondary' amyloid deposits and in certain 'primary' and macroglobulinaemia-related amyloids.

Biophysical and chemical analyses have revealed amyloid to be fibrillar, with typical Congo red staining and polarization birefringence, and an X-ray diffraction pattern characteristic of a β-pleated sheet conformation. Recent detailed X-ray diffraction studies of immunoglobulin subunits have demonstrated a similar conformation in certain parts of the structural domains, including the V_L and C_L domains (see Chapter 8, by Feinstein and Beale). It would seem to be highly significant, therefore, that some amyloid has been shown by immunochemical studies (Glenner et al., 1970) to resemble immunoglobulins in possessing both C and V idiotypic regions (see Pras and Gafni, Chapter 14). Moreover, a comparison of the primary structures of two purified amyloid proteins with those of a well-defined κ Bence–Jones protein (Ker) has provided convincing evidence that the amyloid fibrils are derived from κ light chains (Glenner et al., 1971a), and in particular from the V regions of the latter. But, since amyloid proteins have been found to possess somewhat higher molecular weights than those V_L regions, it has been concluded that they also contain some C-region material or non-immunoglobulin constituents.

The light-chain origin of amyloid deposits is supported also by observations of Glenner and associates (1971b) that precipitates with similar Congo red staining and conformational characteristics are formed when some Bence–Jones proteins are subjected to proteolytic cleavage using pepsin under nearly physiological conditions (0·05 M glycine HC1 at pH 3·5). Moreover, there are reasons to suppose that certain light chains, particularly of the λ subtype, are prone more structurally to form amyloid fibrils. For, the ratio of $\kappa:\lambda$ chains in paraproteins associated with amyloidosis is 1:2 (Pick and Osserman, 1968); and Bence–Jones proteins from patients with amyloidosis have been reported to show a greater tendency to bind to certain normal tissues than do the Bence–Jones proteins from patients with plasma-cell dyscrasias in which amyloidosis has not been observed (Osserman *et al.*, 1964).

Further elucidation of the pathogenesis of amyloidosis could well provide important new information about the aetiology of those paraproteinaemias with which the characteristic deposit formation is particularly associated. Possible origins of amyloid which have been considered so far (see Glenner *et al.*, 1972) include: (*i*) an aberration of a normal light-chain synthetic process, which as suggested by Franklin and Frangione (1971) might involve deletions in the light-chain gene resulting in the production of light-chain fragments analogous to the heavy-chain disease fragments discussed in Section 2.1.4.5.2; and (*ii*) a product of antigen–antibody complex breakdown by macrophages (Waldmann and Strober, 1969).

It is conceivable that further information on the mode of production of amyloid will be provided by studying animal models. In this connection, it should be noted that the susceptibility of mice to amyloidosis correlates neither with the H-2 histocompatibility locus nor with immune traits (Franklin and Clerici, 1974). Current studies on cellular immunological aspects of amyloidogenesis in mice receiving repeated injections of casein should prove rewarding. One noteworthy observation from this approach has been that amyloid and anti-casein antibody production can be elicited in 'nude mice', implying that T cells do not play a prominent role in the genesis of the disease.

2.2 Other Forms of Gammopathy

As mentioned earlier (see Section 1), immunoglobulinopathies have been observed in a whole range of conditions other than malignant immunoproliferative diseases (such as myelomatosis and Waldenström macroglobulinaemia), but usually at relatively lower levels and often in polyclonal forms. Apart from the benign monoclonal gammopathies seen particularly in normal individuals of advancing age (where the level of abnormal protein in the serum is usually less than 2 g/100 ml), gammopathies are also observed in chronic infective states.

These include: (*i*) liver disease, where a particular immunoglobulin class is seen to be predominant in the serum in different forms; for example IgM occurs in primary biliary cirrhosis; IgG in the macronodular chronic aggressive hepatitis; IgA in the micronodular type of Laennec type of cirrhosis (Hobbs, 1971); and IgE in a high incidence of patients with acute or chronic liver disease of widely differing aetiology; (*ii*) Malabsorption states such as active Crohn's disease and ulcerative colitis associated with a raised IgA; (*iii*) skin diseases such as dermatitis herpetiformis, dermatomyositis, and erythema nodosum, in which raised IgA levels are observed; in addition, atopic dermatitis (eczema) is often associated with extremely high levels of serum IgE.

It also should be mentioned that transient paraproteins are observed sometimes (see Young, 1969), which increase rapidly to a peak serum level and disappear spontaneously within a matter of weeks or months. It has been suggested by Hobbs (1971) that such a transient production of paraprotein represents a weak recognition of antigen, but that eventually 'antibody' is produced with a high enough affinity for combination to occur and to switch off the response.

In the context of the present discussions, however, monoclonal gammopathies would seem to be of greater significance. These have been observed, for instance, in a rare chronic skin disease, lichen myxoedematosus; where they always are associated with IgG and, with the absence of Bence–Jones proteinuria (James *et al.*, 1967), and where they are considered to be of benign nature. There is also the possibility that polyclonal gammopathies can, under certain circumstances, convert to monoclonal gammopathy (Osserman and Takatsuki, 1963). This has been observed in animal studies and, possibly, in a young child whose polyclonal pattern turned into a monoclonal one with accompanying Bence–Jones proteinuria (Stoop *et al.*, 1968).

3 IMMUNE DEFICIENCY STATES

3.1 Primary Hypogammaglobulinaemia

The originally recognized immune deficiency state was referred to as agammaglobulinaemia (Bruton, 1952), but later studies employing more sensitive techniques revealed that it was possible to detect some circulating γ-globulin in virtually every case. Hence, the term hypogammaglobulinaemia was adopted. In England, a Medical Research Council Working Party set up to study this condition chose a γ-globulin level of 200 mg/100 ml (100 mg/100 ml for children of less than one year) as the arbitrary level below which such a diagnosis would apply (Soothill, 1962). Immunodeficiency has proved, however, to be considerably more complex than was anticipated at that time.

At the cellular level, primary immunodeficiencies can involve a B-cell defect (as in, for example, infantile sex-linked hypogammaglobulinaemia) or a T-cell defect (as in Di George's syndrome), where there is a congenital absence of the thymus due to defective embryogenesis. In other conditions, both B and T-cell defects occur, as will be seen from Table 10.

Reduced synthesis and secretion of immunoglobulins (IgG, IgA, and IgM) by peripheral blood lymphocytes of patients with common variable hypogammaglobulinaemia were demonstrated when the cells were cultured for several days in the presence of pokeweed mitogen (Waldmann *et al.*, 1974). This has led to the suggestion that this form of immunodeficiency may be attributable

Table 10   Classification of primary immunodeficiency disorders*

Type	Suggested cellular defect		
	B-cells	T-cells	Stem cells
Infantile X-linked agammaglobulinaemia	+		
Selective immunoglobulin deficiency (IgA)	+		
	(some)		
Transient hypogammaglobulinaemia of infancy	+		
X-linked immunodeficiency with hyper-IgM	+	?	
Thymic hypoplasia (pharyngeal pouch syndrome, Di George's syndrome)		+	
Episodic lymphopaenia with lymphocytotoxin		+	
Immunodeficiency with or without hyperimmunoglobulinaemia	+	+	
		(sometimes)	
Immunodeficiency with ataxia telangiectasia	+	+	
Immunodeficiency with thrombocytopaenia and eczema (Wiskott–Aldrich syndrome)	+	+	
Immunodeficiency with thymoma	+	+	
Immunodeficiency with short-limbed dwarfism	+	+	
Immunodeficiency with generalized haematopoietic hypoplasia	+	+	+
Severe combined immunodeficiency			
autosomal recessive	+	+	+
X-linked	+	+	+
sporadic	+	+	+
Variable immunodeficiency (common, largely unclassified)	+	+	
		(sometimes)	

* References to disorders not mentioned above appear in an earlier classification; see *Wld Hlth Org. techn. Rep. Ser.* (1968), No. 402, p. 27
From Bull. Wld Hlth Org. (1971), **45**, 125

to an abnormality of regulatory T cells, which act to suppress B-cell maturation and antibody production. The emphasis in this section will be placed, however, on deficiencies in circulating immunoglobulins.

The first indication of an immunoglobulin deficiency is provided usually by serum electrophoresis, or preferably immunoelectrophoretic analysis (see Figure 10), which is confirmed and measured by a quantitative procedure, using antisera specific for the main immunoglobulin classes and subclasses. In some conditions, such as infantile sex-linked hypogammaglobulinaemia with thymoma and autosomal recessive alymphocytic hypogammaglobulinaemia, all classes of immunoglobulin are extremely deficient. At the other end of the scale, some deficiency states, such as autosomal recessive lymphopenia and thymic aplasia, are characterized by entirely normal circulating immunoglobulin levels. Probably the most interesting cases from the immunochemist's point of view are

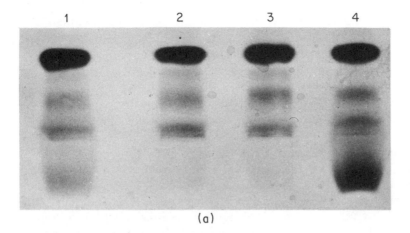

(a)

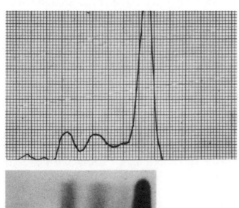

(b)

Figure 10 (a) Cellulose acetate electrophoretic patterns of sera from two patients with primary hypogammaglobulinaemia (2 and 3) compared with normal serum patterns (1 and 4). (b) Densitomeric scan of cellulose acetate electrophoretic strip of hypogammaglobulinaemic serum. (c) Immunoelectrophoretic patterns of hypogammaglobulinaemic serum (1) and normal human serum (2) developed against anti-whole human serum (3) and anti-human IgG (4)

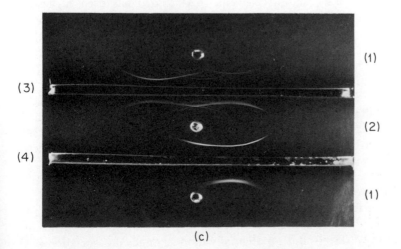

(c)

those showing a selective deficiency of a particular immunoglobin class or subclass. For example some apparently normal individuals have been shown to be deficient in IgA as measured by quantitative immunodiffusion (Rockey *et al.*, 1964). One such case showed classical symptoms of immediate-type hypersensitivity (hayfever). Where such serum IgA deficiency occurs, the two subclasses IgA1 and IgA2 usually are reduced in equal proportions, and there is often a deficiency in the IgA secretions too. Occasionally, however, patients have been encountered with normal levels of circulating IgA but low levels of IgA in secretions and *viceversa* (Goldberg *et al.*, 1969).

There have been reports of selective IgM deficiency in children dying of fulminating meningococcal septicaemia (Hobbs *et al.*, 1967) and in an infant with recurrent *Pseudomomas* infections who possessed IgG and IgA-staining immunofluorescence—but not IgM-staining—plasma cells in the spleen (Faulk *et al.*, 1971). Low levels of IgM have been reported also to be associated with both meningitis and disseminated non-progressive vaccinia (Hobbs *et al.*, 1967).

Selective IgG deficiencies have been observed in patients with recurrent pyogenic infections (Schur *et al.*, 1970), and in some cases deficiencies of particular subclasses have been found to occur. This possibility was suggested originally by the observation of Golebiowska and Rowe (1967) that sera from patients with hypogammaglobulinaemia were sometimes deficient in either slow or fast γ-globulins on immunoelectrophoresis. Since then, various examples of selective IgG deficiency have been reported. For example, IgG2 and IgG3 deficiencies have been observed to be associated with infection in a child (Terry, 1968), and chronic infection has been reported in a family with hereditary deficiency of IgG2 and IgG4 (Oxelius, 1974). Of 59 patients with various immune disorders studied by Yount *et al.* (1970), 11 out of 13 with IgG imbalance showed relative increases in IgG3, especially of the Gm(b) type, and decreases in immunoglobulin with the Gm(g) marker. It is significant, however, that most cases of IgG deficiency so far studied have displayed accompanying deficiencies in other immunoglobulin classes (IgM and IgA), and it has been suggested (Schur, 1970) that the IgG subclasses are linked so closely genetically that it might be impossible to see a deficiency in one subclass which does not affect other subclasses severely and even other classes of immunoglobulin. Indeed, in a recent study of the IgG subclasses in a group of 35 patients with primary hypogammaglobulinaemia all subclasses were found to be depressed in the majority of cases (Shakib, 1976). But an overall comparison of the mean IgG subclass levels of the group, compared with those of normal individuals, revealed that the IgG1 and IgG3 subclasses were least depressed. This finding is consistent with the observed high capacity of the immune system to produce these subclasses not only in young adults, but also in early infancy (Morell *et al.*, 1972; Van der Giessen *et al.*, 1975) and in very old age, that is in individuals over 95-years old (Radl *et al.*, 1975).

It is interesting to note that a patient with ataxia telangectasia with IgA deficiency was found also to be deficient in IgG2, whereas his father was deficient in IgG1 (Rivat *et al.*, 1969). But a potentially more significant association has been suggested between IgA and IgE, on the basis of the observation by Amman and coworkers (1969) of IgE deficiency, as demonstrated by reverse cutaneous anaphylaxis testing with anti-IgE, in 11 out of 16 patients with ataxia telangectasia and IgA deficiency; nine of whom had recurrent sinopulmonary infections of varying degrees. Such observations have been interpreted as evidence that the production of IgA and IgE are linked in some important way. Other findings of IgE deficiency associated with chronic sinopulmonary infection (Cain et al., 1969) have led to the idea that one of the normal functions of IgE is the protection of the respiratory mucosa from infection. A more recent finding of Polmar et al. (1972), that 11 out of 25 individuals with isolated IgA deficiency also lacked circulating IgE (as measured by radioimmunosorbent test), and yet were uniformly asymptomatic, fails to support the concept of a protective role for IgE in respiratory tract immunity. Moreover, a case has been reported (Levy and Chen, 1970) of a healthy individual, with increased incidence of infection, but with no history of allergic disease, who was deficient in circulating IgE; her serum contained less than 0.01 $\mu g/ml$, although her leucocytes did release histamine on incubation with anti-IgE.

Nevertheless, as pointed out by Amman et al. (1970), the immune system is extremely complex in nature and has the ability to compensate for imbalances, so that the absence of any single factor in an apparently normal healthy individual does not necessarily mean that the factor is without a role in defence. Moreover, the high incidence of autoimmune disease and phenomena seen in individuals with selective IgA deficiency (Amman and Hong, 1970) suggests that deficiency of this immunoglobulin class, at least, cannot always be considered to be benign. Other investigators (for example, Koistinen and Sarna, 1975) have made the point that the high incidence of immunological abnormalities in a group of blood donors suggests that IgA deficiency is not a mere laboratory finding.

Another interesting aspect of the question of an interrelationship between IgE and IgA synthesis is the provocative suggestion of Soothill that the onset of atopy of the immediate-type in early life is attributable to a transient IgA deficiency. This is based on the observation by Taylor et al. (1973) that the development of atopy, as indicated by eczema and direct positive skin-test responses to common allergens, in three-month old infants is associated with a serum IgA deficiency when measured by quantitative immunodiffusion. In this connection, it is of possible relevance that approximately one in 700 of a population of mostly healthy individuals have been shown to have a selective IgA deficiency (Bachmann, 1965), and that a familial cluster of such a selective IgA deficiency in a healthy population of blood donors has been reported (Koistinen, 1976). The findings of another study on the serum of IgA levels of 64 588 new blood donors (Koistinen, 1975) bring out the important point that the degree of deficiency of a particular immunoglobulin observed will depend on the accuracy of the assay. For the incidence of IgA deficiency in this large normal population was found to be one in 396 by Ouchterlony analysis (sensitivity: 10 $\mu g/ml$), one in 507 by haemagglutination-inhibition testing (sensitivity: 0.5 $\mu g/ml$), and one in 821 by radioimmunoassay (sensitivity: 0.015 $\mu g/ml$).

The reliability and comparability of the immunoglobulin assays employed is of particular importance in studies of IgE-deficiency states, where it is essential to adopt some radioimmunosorbent procedure. Application of a radiolabelled, modified Mancini method as used by Spitz *et al.* (1972) has failed to detect any measurable IgE (30 ng/ml or more) in the sera of the majority of patients with common variable and sex-linked hypogammaglobulinaemia. Another study (Waldmann *et al.*, 1972) reported that the IgE levels, which had been estimated by using a double-antibody radioimmunoassay were below the limit of detection (4 ng/ml) in nine out of 10 patients with common variable immune deficiency and in seven out of eight patients with thymoma and hypogammaglobulinaemia. On the other hand, a recent study (McLaughlan *et al.*, 1974) has revealed that most patients with common variable and sex-linked hypogammaglobulinaemia had detectable serum IgE (using the radioimmunosorbent procedure). It was also observed that there appeared to be no relationship between the level of IgE in the patients' sera and their reactivity to injected γ-globulin or drugs, or with the occurrence of eczema.

Certain other disorders often are associated with immune deficiency states. For instance, malignancies are particularly frequent in ataxia telangiectasia and in the Wiscott–Aldrich syndrome, where they are responsible for about 10 per cent deaths. There is also a high incidence of autoantibodies, with or without autoimmune disease, in patients with immunodeficiency (as already mentioned see p. 215. In one such example of eight patients with primary hypo-gammaglobulinaemia who developed a severe polyarthritis with some features in common with rheumatoid arthritis, the disease improved dramatically as a result of treatment with γ-globulin (Webster *et al.*, 1976). As classical rheumatoid arthritis does not respond to such treatment, it has been suggested that the arthritis of hypogammaglobulinaemia is due to an organism or toxin which is neutralized by antibody within the injected IgG. In this connection, it is worth noting that an IgA deficiency has been seen to develop occasionally in rheumatoid arthritic patients treated with oral D-penicillamine over relatively long periods. Several such cases have been followed in the author's laboratory (Stanworth *et al.*, 1977). In one the drug was withdrawn immediately when it was realized that the serum level was falling dramatically but it was thought that the subject would have become IgA deficient anyway. Nevertheless, the possibility of drug-induced hypogammopathies of this type in certain genetically predisposed individuals should be recognized.

Conditions which have been reported in association with selective IgA deficiency include recurrent respiratory tract infections, intestinal disorders, juvenile rheumatoid arthritis, and other autoimmune diseases. But, as with the association of polyarthritis referred to above, it would seem more likely that these associated immunological abnormalities are a consequence of the immune deficiency rather than of the genetic defects themselves. As has been pointed out by Good and Yunis (1974) in the primary immune deficiency which develops during ageing in man, and is accompanied by various autoimmune conditions and a high frequency of amyloidosis; antigens and organisms gain access to normally forbidden areas of the body where they persist and abnormally stimulate the residual immunological systems to an excessive degree.

3.2 Secondary Hypogammaglobulinaemia

Hypogammaglobulinaemia can occur sometimes secondary to another disorder, particularly those involving the loss of abnormal amounts of immunoglobulin by excretion (such as occurs in the nephrotic syndrome) or by an aberration of metabolism (such as protein-losing enteropathies like coeliac disease and ulcerative colitis, familial idiopathic hypercatabolic lipoproteinaemia, and myotonic dystrophy). Secondary hypogammaglobulinaemia is substantially more common than the primary form. For instance, secondary deficiency as defined by a serum IgG level of below 200 mg/100 ml, was seen in 0·5 per cent of a group of patients screened at the Hammersmith Hospital, compared with the 10 (0·05 per cent) who proved to be suffering from a primary deficiency (Hobbs, 1968).

It is also interesting to note that polyclonal immunoglobulin of one or other class was found to be reduced significantly in 82 per cent of patients with lymphoproliferative and plasma-cell proliferative disease (Cwynarski and Cohen, 1971), thus accounting for the infection that is frequently the terminal event in such cases. Immunoglobulin deficiency was found to be equally prevalent in all classes of monoclonal gammopathy, but was considerably more common in myelomatosis than in benign monoclonal gammopathy. Moreover, evidence has been obtained from cell culture experiments, which suggests that the observed deficiencies in circulating polyclonal immunoglobulins in these conditions are attributable to a failure of synthesis. Polyclonal deficiency in macroglobulinaemia is ascribed to the peripheral spread of neoplastic cells, a process which apparently does not account for the immunoglobulin deficiencies observed in typical cases of myelomatosis.

With regard to the association of hypogammaglobulinaemia with renal disease, it is interesting to note that an asymmetric depression has been observed in the serum levels of certain IgG subclasses in some patients with 'minimal-change' nephrotic syndrome, focal glomerulosclerosis, and proliferative glomerulonephritis (Shakib et al., 1976). This suggests that the urinary loss of IgG alone cannot account for the low blood levels of this immunoglobulin class encountered in these conditions.

4 EXPERIMENTAL MODELS

4.1 Experimental Paraproteinopathies

4.1.1 Induction of Hypergammaglobulinaemia in Animals

Although, as far as is known, naturally occurring or experimentally induced plasma-cell neoplasms have not been reported in rabbits, as they have in other species such as mice, the experimental immunization of this species with

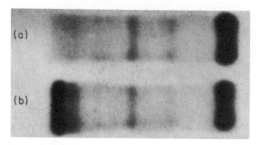

Figure 11 Zone electrophoretic patterns of serum from a rabbit taken (a) before and (b) after immunization with group A streptococcal vaccine. Reproduced from Osterland *et al.* (1966), with permission

streptococcal carbohydrates has led *in some animals* to the production of an electrophoretically uniform γ-globulin at an elevated serum concentration of the order of 20–50 mg/ml. This, in many ways, resembles the monoclonal paraproteins seen in the sera of patients with multiple myeloma (Osterland *et al.*, 1966). Figure 11 shows microzone electrophoretic patterns of the serum of rabbits before and four weeks after intravenous immunization with group A-variant *Streptococcus*. Similar homogeneous antibody responses have been observed in some rabbits immunized with types III and VIII pneumococcal capsular polysaccharide (Pincus *et al.*, 1970a, b).

Apart from the zone electrophoretic homogeneity, evidence for restricted heterogeneity of such antibodies has been indicated by: (*i*) the monodisperse distribution of their light chains on disc electrophoresis, (*ii*) allotype exclusion, and (*iii*) the identical amino acid sequences of the N-terminal regions (residues 1–41) of their light chains (Braun *et al.*, 1975). In all respects, therefore, the rabbit antibodies appear to be as homogeneous as human myeloma immunoglobulin.

Hopefully, the study of this animal model will provide important information about those malignant processes in humans that lead to the production of monoclonal immunoglobulins. It is interesting to note, therefore, that of the various factors which have been found to influence the production of paraproteinaemia in rabbits, the most important appears to be the choice of animal with the appropriate genetic background and the repeated intravenous immunization with a vaccine composed of whole bacteria, with intact surface carbohydrate (Braun *et al.*, 1969). It has been found, for instance, that immunization with vaccine *via* the intraperitoneal route is much less effective, as is the use of purified soluble bacterial carbohydrates alone as immunogens. In addition, the selective breeding of high and low-responding rabbits has provided data which suggest that the magnitude of the immune response is transmitted genetically.

Histological examination of rabbits sacrificed at the time of maximal serum paraprotein level has revealed intensive proliferation of plasma cells in the spleen, lungs, and lymph nodes (Braun and Krause, 1969). Immunofluorescence studies showed that the majority of plasma cells in the spleens of rabbits immunized with type VII *Pneumoccocus* species were indeed synthesizing antibodies to this antigen. It has been suggested by Krause (1970) that the intense immunization by repeated intravenous injections of vaccine over a prolonged period selects a restricted population of cells which undergoes proliferation. Attention has been drawn to an interesting parallelism with the development of a myeloma-like condition in mink with Aleutian disease. Late in the course of this condition some mink show a transition from a heterogeneous hypergammaglobulinaemia to a homogeneous myeloma-like disease, suggesting the ascendancy of a few predominant clones of plasma cells (Porter *et al.*, 1965). As mentioned earlier, it is conceivable that a similar transition sometimes occurs in those humans who initially exhibit a benign polyclonal hypergammaglobulinaemia which ultimately develops into a malignant monoclonal form.

As in the human condition, certain rabbits in which paraproteinaemia has been induced experimentally by immunization with bacterial vaccines exhibit cryoglobulinaemia. For instance, cryoglobulin production in New Zealand Red rabbits immunized with streptococcal (group B/C) vaccines began around three weeks after the start of immunization and reached maximal serum levels ranging from less than 0·1 mg/ml to greater than 6 mg/ml (Herd, 1973b). It should be noted that these experimentally induced cryoglobulins were shown to comprise IgM and IgG—including IgM anti-γ-globulin and homogeneous IgG anti-streptococcal antibodies—as well as DNA. Hence, they resemble closely the composition of the spontaneously occurring human counterparts and, for this reason, would seem to offer a useful model. In this connection, Herd (1973b) has postulated, on the basis of her studies using the experimental rabbit system, that cryoglobulins are formed in two stages, the first involving structural modification of IgG followed by its precipitation with anti-γ-globulin. She has also suggested (Herd 1973a) that anti-γ-globulins might be operative in the experimental production of paraproteinaemia in rabbits, in suppressing immunoglobulin synthesis by heterogeneous clones of antibody-forming cell precursors while stimulating proliferation of selected clones which secrete the homogeneous antibodies. Evidence in support of this idea has been provided from recent studies of antibody production in rabbits immunized with streptococcal group A and C carbohydrate antigens (Aasted, 1974), which have revealed that the appearance of an antibody response of restricted heterogeneity is accompanied by the production of anti-antibody (IgG) as measured by the latex fixation test. It is possible that such anti-antibodies are directed against previously secluded determinants within the Fc region of the original anti-streptococcal IgG antibody, which are revealed because of a conformational change brought about by combination with the bacterial antigen. This probably occurs in a manner similar to that postulated to explain the production of anti-γ antibodies in rabbits immunized with preformed antigen–antibody (rabbit IgG) complexes (Henney and Stanworth, 1966). In this connection, it is significant that 7 S anti-IgG antibodies have been isolated by immunosorption from the sera of rabbits immunized against streptococcal carbohydrates which have specificity for the Fc region of rabbit IgG (Bokisch *et al.*, 1973), and which are themselves also of restricted heterogeneity, besides possessing idiotypic specificity.

4.1.2 Spontaneous or Experimentally Induced Plasma-Cell Tumours

The study of hyperimmunoglobulinopathy has also been facilitated by the availability of spontaneously occurring or experimentally induced plasma-cell tumours in species such as rats and mice. For instance, the inbred C3H (Potter *et al.*, 1957) and the F_1 hybrids of CBA $\times$ DBA/2 mice (Rask-Neilsen *et al.*, 1959) show a unique predilection to develop 'spontaneous' plasma-cell tumours. Consequently, these animals have been used widely in the study of genetic and other factors relevant to the pathogenesis of plasma-cell neoplasia. Another example of a highly valuable experimental model is the spontaneously occurring ileocoecal immunocytoma frequently found in the inbred LOU/WS1 strain of rat, whose secreting properties are maintained over many passages in histocompatible animals (Bazin *et al.*, 1974). Such tumours have been shown to secrete monoclonal immunoglobulin which possesses immunological, chemical, and biological properties characteristic of rat IgE.

Alternatively, immunoglobulin-secreting tumours have been induced experimentally in various species by transplantation, or by injection, of some suitable irritant. For example, one transplantation line of leukaemia in the (CBA $\times$ DBA/2) F_1 murine strain is characterized by the appearance of a macroglobulin component in the serum, and a histological pattern resembling human macroglobulinaemia (Clausen *et al.*, 1960). Another, a transplanted plasma-cell tumour of ileococcal origin ($\times$ 5563) in the C3H strain, has been shown to be histologically similar to the neoplasms of human multiple myeloma and to be accompanied by a similar development of osteolytic bone lesions and the production of an abnormal serum protein of γ electrophoretic mobility, which is also extractable from the tumours (Potter *et al.*, 1957).

Plasma-cell neoplasms also have been induced experimentally in strain BALB/c mice by intraperitoneal injection of a mineral oil–*Staphylococcus* mixture, or even mineral oil alone. The highest incidence of tumours was observed in mice receiving three injections (0·5 ml) of the latter at two-months intervals (Potter and Boyce, 1962). Similar intraperitoneal plasma-cell neoplasms have been induced in this mouse strain by implantation of a Millipore diffusion chamber (Merwin and Algire, 1959); it being concluded that, like mineral oil, this is only mildly irritating and therefore tolerated well by most mice, and is not removable by host defence mechanisms but can cause a reactive tissue on the peritoneal surfaces.

As mentioned earlier, a marked hypergammaglobulinaemia is a regular feature of virus-inducible Aleutian disease in mink (see Section 4.1.1); which shows histological features in common with human plasma-cell dyscrasias, such as a parallelism between the severity of the disease and the rise in serum levels of 7 S γ-globulin (Henson *et al.*, 1963). Moreover, the hypergammaglobulinaemia as revealed by paper electrophoretic analysis of serum in some cases transforms into a homogeneous myeloma-like disease with time, resulting in the appearance of Bence–Jones protein in the urine (Porter *et al.*, 1965). This has been interpreted as being due to the transition from a hyperplastic state, where the viral agent is assumed to be affecting a number of clones of plasma cells, to a neoplastic one involving monoclonal plasma-cell proliferation.

4.1.3 In Vitro *Culture of Paraprotein-Producing Cell Lines*

The *in vitro* synthesis and secretion of immunoglobulins by myeloma and lymphoblastoid-cultured cell lines is now well established (Table 11), and has already provided valuable information about the factors influencing paraprotein production in the patient, besides offering an alternative source of monoclonal immunoglobulin. For instance, the immunoglobulin produced by established cell lines of human multiple myeloma origin derived from buffy coat material was of a single light-chain type and showed well-defined electrophoretic mobility (Matsuoka *et al.*, 1968). Thus, one cell line (RPM1 8235) produced only κ-type IgG, another line (RPM1 4666) produced κ-type IgA and free κ light chains, while a third (RPM1 8226) produced only λ light chains. Moreover, the time course of immunoglobulin accumulation and cell proliferation indicates that the cells were able to synthesize and secrete immunoglobulin actively only while they were able to proliferate.

One of these cell lines (RPM1 8226) has been cultured continuously over a number of years, and been shown to continue to grow well and to produce only λ-

Table 11   Ig-secreting cell lines of human haemapoietic origin

Number	Source	Characteristics of Ig product(s)	Secretion rate	Reference
RPMI 8235	Buffy coat of patients with chronic myelogeneous leukaemia	IgG (κ) only	3 mg Ig/10^6 cells day^{-1}*	
RPMI 4666	,,	IgA (κ); free κ light chains (inbalance of heavy and light-chain synthesis)	,, +	Matsuoka *et al.* (1968)
RPMI 8226	Buffy coat of patient with multiple myeloma	Free light chains only	220 ng light chains/10^6 cells day^{-1}*	
266 B1,	Bone marrow from patient with multiple myeloma	IgE	$8 \cdot 1 \times 10^{-12}$ g†/ cell hr^{-1}	Nilsson (1971)

* In most active phases of synthesis and secretion   † Max. rate of synthesis.

type light chains (Matsuoka *et al.*, 1969). Moreover, many of the cells were seen to resemble morphologically immature plasma cells with well-developed endoplasmic reticula and numerous ribosomes. Furthermore, the light-chain material isolated from these cultures was shown to possess physical and chemical features characteristic of λ-type Bence–Jones proteins, sedimenting as a dimer ($S^{\circ}_{20,\omega} = 3\cdot6$ at 12·5 mg/ml) at neutral pH in the ultracentrifuge (Matsuoka *et al.*, 1969).

IgE-producing cell lines (266B1 and 268 Bm) also have been established *in vitro* from the first case of human myeloma of this type, reported in Sweden (Nilsson *et al.*, 1970; Nilsson, 1971). From this work it was concluded that myeloma cells are unusually exacting in their metabolic requirements, and only if media rich in vitamins and amino acids (for example F10 and RPM1 1640) are conditioned by the presence of growing cells will the *in vitro* IgE production be optimal. It will then approach the secretion rate obtained in fresh explants of human myeloma cells, which is presumed to reflect the *in vivo* synthetic rate of immunoglobulins. Indeed, the average rate of synthesis of IgG by myeloma cells *in vitro* has been used, together with a quantitative measurement of the total body rate of IgG synthesis *in vivo*, to estimate the numbers of tumour cells in patients with IgG myeloma (Salmon and Smith, 1970) by substitution in the equation:

$$MCT_{\mathrm{B}} = \frac{RT_{\mathrm{B}}}{R_{\mathrm{M}}} \times MC\ i.v.$$

where MCT_{B} = total body myeloma-cell number, $MC\ i.v.$ = number of myeloma cells *in vitro*, RT_{B} = rate of total body IgG synthesis (g/24 hr) *in vivo* R_{M} = rate of total myeloma IgG synthesis (g/24 hr) *in vitro*. The tumour cell number thus determined was then used in the clinical evaluation of the extent of the disease, being employed to estimate the average rate of growth of this particular malignancy.

Average molecular synthetic rates in different patients have been found to range from 12 500 to 81 000 molecules IgG/myeloma cell min^{-1}; and the total body myeloma-cell number was found to range from $0\cdot5 \times 10^{12}$ to $3\cdot1 \times 10^{12}$ myeloma cells, and could be correlated with the degree of skeletal damage observed on radiographs. Likewise, a patient with IgE myelomatosis (P.S in Table 2) was shown to synthesize IgE molecules at a comparable rate of 26 000 molecules/myeloma cell min^{-1}, despite a concomitant plasma-cell leukaemia (Salmon *et al.*, 1971). However, autoradiographic studies demonstrated a significantly greater labelling of DNA in this patient's bone marrow cells than is seen in cells from IgG-myeloma patients.

The IgE-producing cell line 266B1 has continuously produced, over a long period, immunoglobulin molecules identical to those synthesized by the patient *in vivo* at an optimal rate over a 48-hour period of $1\cdot7 \times 10^{-13}$ g IgE/cell hr^{-1}. The maximal rate of synthesis ($8\cdot1 \times 10^{-12}$ g IgE/cell hr^{-1}) was correlated with rapid cell growth (at 10^{6}/30 ml cell densities), and with the presence of feeder human skin fibroblasts or glia-like cells, or with the use of conditioned media harvested from such cells.

The study of *in vitro* systems is providing valuable information about the regulation of synthesis and secretion of immunoglobulins by myeloma cells. An *in vitro* approach, which is potentially even more powerful is based initially on the isolation of mutant mouse myeloma cell lines obtained by selecting single cells from a continuous culture of mouse MOPC 21 myeloma cells (Cotton *et al.*, 1973; Secher *et al.*, 1973). Four immunoglobulin structural mutants produced spontaneously by such cell lines have been characterized using amino acid sequencing, cell-free synthesis, and studies on the isolated mRNA, and have been shown to arise due to defects in the heavy-chain cistron (Milstein *et al.*, 1977). Thus, mutant proteins (IF1 and IF3) involve deletions of the C_H3 domain (see Figure 12), due to early termination and to a frame shift, respectively. IF2 has an internal deletion resembling the heavy-chain disease of humans and IF4 (not illustrated in Figure 12) shows an amino acid substitution due to point mutation.

An even more exciting advance in this area, however, has been the development by cell fusion of new lines capable of antibody synthesis. This has been accomplished by, for example, fusing mouse myeloma cells with spleen cells from normal mice which have been immunized with sheep erythrocytes or hapten-carrier antigens, using TNP as the hapten. The resultant hybrids express both the parental myeloma polypeptide chains and, in addition, a light and heavy chain responsible for antibody specificity. Methods for the specific detection of clonal variants have been devised, and it has been shown that such clonal lines can be grown as tumours in mice, from whose sera specific antibody can be isolated.

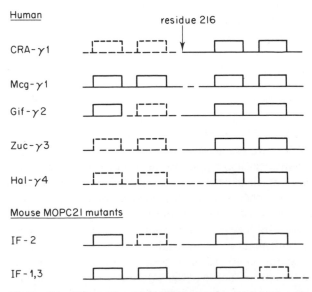

Figure 12 Comparison of deletions observed in human heavy chains and in mouse mutants. Reproduced from Milstein *et al.* (1974) with permission

4.2 Experimental Hypogammaglobulinaemia

As in the experimental study of hypergammaglobulinaemia (see Sections 4.1.1 and 4.1.2), the growing use of animal models is providing a valuable insight into the various factors influencing the induction and perpetuation of immuno-deficiency states in humans, as well as pointing to possible means of clinical treatment. It is thus possible to investigate both the cellular and immune elements of immunodeficiency states, and their underlying genetic defects.

The three main approaches which have been adopted concern: (*i*) congenital, and surgically or chemically induced defects, involving the thymus, or the bursa of Fabricius in chickens, (*ii*) infectious unresponsiveness, and (*iii*) allotypic suppression. Examples of each will be considered.

The study of congenitally thymus-lacking so-called 'nude' (nu/nu) mice offers a means of determining the influence of T cells on immunoglobulin synthesis. For instance, by this approach it has been shown that the establishment of normal serum concentrations of IgA, IgG2a, and especially IgG1, in this species requires a viable thymus; whereas IgM production shows no such T-cell dependency (Pritchard *et al.*, 1973). On the other hand, studies in chickens which have been 'bursectomized' hormonally suggest that impaired J-chain synthesis may constitute a limiting process in IgM production by B cells in this species (Ivanyi, 1975). Other congenital deficiencies in animals, such as the fatal one seen in arab foals, involve a combined B and T-lymphocyte impairment (McGuire *et al.*, 1974).

Loss of bursal function also can arise in chickens as a result of naturally occurring viral infections. An example of such a situation is infectious bursal disease, formerly known as Gumboro disease, which is characterized by the destruction of the lymphoid tissue in the bursa without repopulation (Cheville, 1967). The observed reduction in haemagglutinating antibody (IgG) response to Newcastle disease vaccine and in the serum IgG levels of such birds has been interpreted as providing further evidence that the switch over from IgM to IgG antibody production takes place within the bursa (Faragher *et al.*, 1972). The observation is also consistent with an earlier report that bursectomy in late embryonic life has no influence on IgM antibody responses in chickens, but prevents an IgG response.

The third experimental approach to the induction of immunodeficiency states in animals involves the suppression of synthesis of immunoglobulin molecules carrying particular allotypic markers by treatment with anti-allotype antiserum. Long-lasting suppression can be achieved with relative ease and regularity in the rabbit in this way. In the mouse suppression is obtained only with certain strain combinations and tends to be sporadic (Adler, 1975). Although the type of allotype suppression first described in the rabbit by Dray (1962) was brought about by maternal transmission of the anti-allotype (isoantibody), suppression is achieved most conveniently in experimental models by neonatal injection of antiserum (Mage and Dray, 1965). Suppression thus induced can be directed selectively and specifically against alleles of the two genetic loci used by cells in the synthesis of all classes of rabbit immunoglobulin: the a locus (on the Fd region of the heavy chain) and the b locus (on κ light chains). Possible mechanisms of this form of allotypic suppression have been proposed recently by Catty *et al.* (1975). Cultured spleen cells from rabbits which have been chronically suppressed in this way provide a useful *in vitro* model. Adler (1975) has shown that treatment of these cells with anti-allotypic serum specific for the non-suppressed allotype will abrogate the experimentally induced suppression. This has provided a valuable means of learning more about the suppression mechanism, by studying those conditions under which such reversal of suppression occurs.

5 PARAPROTEINS AS TOOLS IN IMMUNOGLOBULIN CHARACTERIZATION

The availability of monoclonal paraproteins representative of the various classes and subclasses has advanced the structural elucidation of human immunoglobulins. This is particularly the case, of course, with regard to IgE and IgD, which normally are found at such low levels in the sera of non-neoplastic individuals that their isolation in sufficient amounts for full characterization has presented a most formidable task. Heavy polypeptide chains of paraproteins of each human class and subclass have been or have are being sequenced, following the primary structural elucidation of light-chain (including Bence–Jones protein) preparations. Where crystalline forms are available, tertiary structural studies are also in progress. Moreover, the paraproteins produced by experimental animals such as those of ileococcal origin in the rat (mentioned in Section 4.1.2), likewise offer the key to the structural characterization of immunoglobulins of these species; as do the highly homogeneous antibodies produced by some rabbits immunized with streptococcal vaccines (see Section 4.1.1).

Such paraproteins are also proving valuable tools in functional studies of their normal immunoglobulin counterparts. For instance, proteolytic and chemical-cleaved fragments of the first described myeloma IgE protein were used in inhibition of Prausnitz-Küstner (PK) (Stanworth *et al.*, 1967, 1968) and inhibition of passive cutaneous anaphylaxis (PCA) (Stanworth, 1973*b*) tests to provide convincing evidence that reaginic antibodies belong to this new class of immunoglobulin and that they bind to target mast cells through sites within their Fc regions. Likewise studies with the rat myeloma IgE protein (Bazin *et al.*, 1974), mentioned in Section 4.1.2, have confirmed that rat reaginic antibody molecules are also of this class. To quote another example of this type, chemical cleavage fragments of the mouse myeloma protein IgG2a MOPC 173 have been employed in investigations aimed at locating the position of complement-activating sites (Kehoe *et al.*, 1969). Similarly, inhibition studies with human myeloma proteins have demonstrated the association of certain biological activities with particular IgG subclass; for example this approach has been employed in studies of the inhibition of monocyte phagocytosis of erythrocytes coated with anti-Rh antibodies to show that monocyte binding is restricted to the IgG1 and IgG3 subclasses (Huber and Fudenberg, 1968). There are many analogous examples, involving studies of the inhibition of other Fc-located biological activities by certain myeloma immunoglobulin subclasses and their cleavage fragments (see Stanworth and Turner, 1973; Stanworth, 1974; Stanworth and Stewart, 1975).

An alternative, but parallel, approach has involved the use of incomplete or deleted paraproteins in these sort of tests for Fc-located immunoglobulin functions. An example of this approach is the demonstration of the failure of IgG proteins lacking a C_H3 domain (IF1, IF3), obtained from mutant cell lines of MOPC/2 mice, to inhibit rosette formation by mouse lymphocytes, in contrast to another mutant protein (IF2) which lacks the C_H1 domain. This provides further evidence that the C_H3 region is essential for the binding of IgG to Fc receptors on lymph-node cells (Ramasamy et al., 1975). The further study of paraproteins with more limited deletions could well help to pinpoint more precisely the location of various cell-binding and other Fc-located effector sites.

Conversely, the studies of hapten-binding paraproteins (referred to earlier in Section 2.1.4.4) have provided important information about the nature of the Fab-located antigen-binding sites of antibody molecules. Thus, the gammopathies, whether occurring spontaneously or induced experimentally, have presented and will continue to present a rich source of material for the immunochemist interested in the nature and role of immunoglobulins in health and disease.

ACKNOWLEDGEMENTS

The author is most grateful to his colleagues Philip Johns and Helen Evans for providing the illustrations reproduced in Figures 2, 3, 5 and 10; and to the other people who have permitted him to reproduce their data.

REFERENCES

Aasted, B. (1974). *Scand. J. Immunol.*, **3**, 553.

Adler, L. T. (1975). Transplant. Rev., **27**, 3.

Alper, C. A. (1966). *Acta med. scand.*, **Suppl. 445,** 200.

Amman, A. J., Cain, W. P., Ishizika, K., Hong, R., and Good, R. A. (1969). *New Eng. J. Med.*, **281**, 469.

Amman, A. J., and Hong, R. (1970). *Clin. Exp. Immunol.*, **7**, 883.

Amman, A. J. (1970). *New Eng. J. Med.*, **283**, 542.

Apitz, K. (1940). *Virchows Archs Path. Anat.*, **306**, 631.

Asquith, P., Thompson, R. A., and Cooke, W. T. (1969). *Lancet*, **ii**, 129.

Auscher, C., and Guinand, S. (1964). *Clin. Chim. Acta*, **9**, 40.

Axelsson, U., and Bachmann, R., and Hällén, J. (1966). *Acta med. scand.*, **179**, 235.

Bachmann, R. (1965). *Scand. J. Clin. Lab. Invest.*, **17**, 316.

Ballieux, R. E., Imhof, J. W., Mul, N. A. J., Zegers, B. J. M., and Stoop, J. W. (1968). *Clin. Chim. Acta*, **22**, 7.

Bauer, K. (1974). In *Progress in Immunology II*, Vol. **3**, (Brent, L., and Holborow, J., eds.), North Holland Publishing Co., Amsterdam, p. 390.

Bazin, H., Querinjean, P., Becker, S. A., Heremans, J. F., and Dessy, F. (1974). *Immunology*, **26**, 713.

Bennich, H., and Johansson, S. G. O. (1967). *Immunology*, **13**, 381.
Bernier, G. M., Ballieux, R. E., Tominaga, K., and Putnam, F. E. (1967). *J. Exp. Med.*, **125**, 303.
Bessis, M., Briton-Gorius, J., and Binet, J. L. (1963). *Nouv. Rev. Franc. Hematol.*, **3**, 159.
Bokisch, V. A., Chia, O. J. W., and Bernstein, D. (1973). *J. Exp. Med.*, **137**, 1354.
Bradley, J., Hawkins, C. F., Rowe, D. S., and Stanworth, D. R. (1968). *Gut*, **9**, 564.
Braun, D. G., Eichmann, K., and Krause, R. M. (1969). *J. Exp. Med.*, **129**, 809.
Braun, D. G., Huser, H., and Jaton, J.-C. (1975). *Nature*, (*Lond.*), **258**, 363.
Braun, D. G., and Krause, R. M. (1969). *Z. Immun. Forsch.*, **139**, 104.
Bruton, O. C. (1952). *Pediatrics*, **9**, 722.
Buxbaum, J. (1973). *Semin. Hematol.*, **10**, 33.

Cain, W. A., Ammann, A. J., Hong, R., Ishizaka, K., and Good, R. A. (1969). *J. Clin. Invest.*, **48**, 120.
Capra, J. D., and Kunkel, H. G. (1970). *J. Clin. Invest.*, **49**, 610.
Catty, D., Lowe, J. A., and Gell, P. G. H. (1975). *Transplant. Rev.*, **27**, 157.
Cheville, N. F. (1967). *Am. J. Pathol.*, **51**, 527.
Cioli, D., and Baglioni, C. (1966). *J. Mol. Biol.*, **15**, 385.
Clausen, J., Rask-Nielsen, R., Christiensen, H. E., Lontie, R., and Herremans, J. (1960). *Proc. Soc. Exp. Biol. Med.*, **103**, 802.
Cotton, R. G. H., Secher, D. S., and Milstein, C. (1973). *Eur. J. Immunol.*, **3**, 135.
Cwynarski, M. T., and Cohen, S. (1971). *Clin. Exp. Immunol.*, **8**, 237.

Dray, S. (1962). *Nature*, (*Lond.*), **195**, 677.

Faragher, J. T., Alan, W. H., and Cullen, G. A. (1972). *Nature, New Biol.*, **237**, 118.
Faulk, W. P., Kiyasu, W. S., Cooper, M. D., and Fudenberg, H. H. (1971). *Pediatrics*, **47**, 399.
Feizi, T. (1967). *Science, N.Y.*, **156**, 1111.
Fishkin, B. G., Orloff, N., Scaduto, L. E., Borucki, D. T., and Spiegelberg, H. L. (1972). *Blood*, **39**, 361.
Florin-Christensen, A. (1974). *Curr. Titles*, **2**, (11), p. 275.
Florin-Christensen, A., Donaich, D., and Newcomb, P. J. (1972). *Brit. Med. J.*, **ii**, 413.
Florin-Christensen, A., and Roux, R. (1974). In *Progress in Immunology II*, Vol. I, (Brent, L., and Holborow, J., eds.), North Holland Publishing Co., Amsterdam, p. 285.
Frangione, B., and Franklin, E. C. (1973). *Semin. Hematol.*, **10**, 53.
Franklin, E. C. (1964). *J. Exp. Med.*, **120**, 691.
Franklin, E. C. (1975). *Archs Int. Med.*, **135**, 71.
Franklin, E. C., and Clerici, E. (1974). In *Progress in Immunology II*, Vol. 4, (Brent, L., and Holborow, J., eds.), North Holland Publishing Co., Amsterdam, p. 381.
Franklin, E. C., and Frangione, B. (1971). *Proc. Nat. Acad. Sci., U.S.A.*, **68**, 187.
Franklin, E. C., and Frangione, B. (1975). *Contemp. Top. Mol. Biol.*, **4**, 89.
Franklin, E. C., Meltzer, M., Guggenheim, F., and Lowenstein, J. (1963). *Fed. Proc.*, **22**, 264.
Freel, R. J., Maldonado, J. E., and Gleich, G. J. (1972). *Am. J. Med.*, **264**, 117.
Fudenberg, H. H. (1972). In *Cancer Chemotherapy II* (Brodsy, I., Kahn, S. B., and Moyer, J. H., eds.), Grune and Stratton Inc., New York, p. 393.

Glenner, G. G., Ein, D., Eanes, E. D., Bladon, H. A., Terry, W., and Page, D. (1971*b*). *Science, N.Y.*, **174**, 712.
Glenner, G. G., Ein, D., and Terry, W. D. (1972). *Am. J. Med.*, **52**, 141.
Glenner, G. G., Harada, M., Isersky, C., Cuatrecasas, C., Page, D., and Keiser, H. R. (1970). *Biochem. Biophys. Res. Commun.*, **41**, 1013.

228

Glenner, G. G., Terry, W., Harada, M., Isersky, C., and Page, D. (1971a). *Science, N.Y.*, **171**, 1150.
Goldberg, L. S., and Barnett, E. V. (1970). *Ann. Int. Med.*, **125**, 145.
Goldberg, L. S., Douglas, S. S., and Fudenberg, H. H. (1969). *Clin. Exp. Immunol.*, **4**, 579.
Golebiowska, H., and Rowe, D. S. (1967). *Clin. Exp. Immunol.*, **2**, 275.
Good, R. A., and Yunis, E. (1974). *Fed. Proc.*, **33**, 2040.
Grey, H. M., and Kunkel, H. G. (1964). *J. Exp. Med.*, **120**, 253.

Hallén, J. (1963). *Acta med. scand.*, **173**, 737.
Harboe, M., and Feizi, T. (1974). In *Progress in Immunology II*, Vol. 3, (Brent, L., and Holborow, J., eds.), North Holland Publishing Co., Amsterdam, p. 390.
Harboe, M., and Lind, K. (1966). *Scand. J. Haematol.*, **3**, 269.
Harboe, M., van Furth, R., Schubothe, H., Lind, K., and Evans, R. S. (1965). *Scand. J. Haematol.*, **2**, 259.
Henney, C. S., and Stanworth, D. R. (1966). *Nature, (Lond.)*, **210**, 1071.
Henson, J. B., Gorham, J. R., and Leader, R. W. (1963). *Nature, (Lond.)*, **197**, 206.
Herd, Z. L. (1973a). *Immunology*, **25**, 923.
Herd, Z. L. (1973b). *Immunology*, **25**, 931.
Hobbs, J. R. (1968). *Proc. Roy. Soc. Med.*, **61**, 883.
Hobbs, J. R. (1969). *Brit. J. Haematol.*, **16**, 599.
Hobbs, J. R. (1971). *Adv. Clin. Chem.*, **14**, 219.
Hobbs, J. R., Milner, R. D. G., and Watt, P. J. (1967). *Brit. Med. J.*, **iv**, 583.
Huber, H., and Fudenberg, H. H. (1968). *Int. Archs Allergy*, **34**, 18.

Isobe, T., and Osserman, E. F. (1974). *Blood*, **43**, 505.
Ivanyi, I. (1975). *Immunology*, **28**, 1015.

James, K., Fudenberg, H., Epstein, W. L., and Schuster, J. (1967). *Clin. Exp. Immunol.*, **2**, 153.
Jancelwicz, Z., Takatuski, K., Sugai, S., and Pruzanski, W. (1975). *Archs Int. Med.*, **135**, 87.

Kehoe, J. M., Fougereau, M., and Bourgois, A. (1969). *Nature, (Lond.)*, **224**, 1212.
Kiss, M., Fudenberg, H. H., and Kritzman, J. (1967). *Clin. Exp. Immunol.*, **2**, 467.
Knedel, M., Fateh-Moghadam, A., Edel, H., Bartl, R., and Neumeier, D. (1975). *Dtsch. Med. Woch.*, **101**, 496.
Koistinen, J. (1975). *Vox. Sang.*, **29**, 192.
Koistinen, J. (1976). *Vox. Sang.*, **30**, 181.
Koistinen, J., and Sarna, S. (1975). *Vox. Sang.*, **29**, 203.
Krause, R. M. (1970). *Adv. Immunol.*, **12**, 1.
Kunkel, H. G., Natvig, J. B., and Joslin, F. G. (1969). *Proc. Nat. Acad. Sci., U.S.A.*, **62**, 144.

Leddy, J. P., Deitchman, J., and Bakemeier, R. F. (1970). *Arth. Rheumat.*, **13**, 331.
Levy, D. A., and Chen, J. (1970). *New Engl. J. Med.*, **283**, 541.

Mage, R. G., and Dray, S. (1965). *J. Immunol.*, **95**, 525.
Martin, N. H. (1970). *Brit. J. Hosp. Med.*, **3**, 662.
Matsuoka, Y., Takahashi, M., Yagi, Y., Moore, G. E., and Pressman, D. (1968). *J. Immunol.*, **101**, 1111.
Matusoka, Y., Yagi, Y., Moore, G. E., and Pressman, D. (1969). *J. Immunol.*, **102**, 1136.
McGuire, T. C. *et al.* (1974). *J. Am. Vet. Med. Ass.*, **164**, 70.
McKenzie, M. R., Fudenberg, H. H., and O'Reilly, R. A. (1970). *J. Clin. Invest.*, **49**, 15.

McLaughlan, P., Stanworth, D. R., Webster, A. D. A., and Asherson, G. L. (1974). *Clin. Exp. Immunol.*, **16**, 375.

Meltzer, M., and Franklin, E. C. (1966). *Am. J. Med.*, **40**, 828.

Meltzer, M., Franklin, E. C., Elias, K., McCluskey, R. T., and Cooper, N. (1966). *Am. J. Med.*, **40**, 837.

Merwin, R. M., and Algire, G. H. (1959). *Proc. Soc. Exp. Biol. Med.*, **101**, 437.

Mills, R. J., Fahie-Wilson, M. N., Carter, P. M., and Hobbs, J. R. (1976). *Clin. Exp. Immunol.*, **23**, 228.

Milstein, C., Adetugbo, K., Cowan, N. J., Köhler, G., Secher, D. S., and Wilde, C. D. (1977). *Cold Spring Harbor Symposium on Quantitative Biology*, **41**, 793.

Milstein, C., Adetugbo, K., Cowan, N. J., and Secher, D. S. (1974). In *Progress in Immunology II*, Vol. I, (Brent, L., and Holborow, J., eds.), North Holland Publishing Co., Amsterdam, p. 157.

Morell, A., Skvaril, F., Hitzig, W. H., and Barandum, S. (1972). *J. Pediatrics*, **80**, 960.

Moroz, C., Amir, J., and de Vries, A. (1971). *J. Clin. Invest.*, **50**, 2726.

Natvig, J. B., Kunkel, H. G., and Litwin, S. D. (1967). *Cold Spring Harbor Symp. Quant. Biol.*, **32**, 173.

Nilsson, K. (1971). *Int. J. Cancer*, **7**, 380.

Nilsson, K., Bennich, H., Johansson, S. G. O., and Pontén, J. (1970). *Clin. Exp. Immunol.*, **7**, 477.

Ogawa, M., Kochwa, S., Smith, C., Ishizaka, K., and McIntyre, O. R. (1969). *New Engl. J. Med.*, **281**, 1217.

Osserman, E. F., and Kohn, J. (1974). In *Progress in Immunology II*, Vol. 3, (Brent, L., and Holborow, I., eds.), North Holland Publishing Co., Amsterdam, p. 390.

Osserman, E. F., and Takatsuki, K. (1963). *Medicine*, **42**, 357.

Osserman, E. F., and Takatsuki, K. (1964). *Am. J. Med.*, **37**, 351.

Osserman, E. F., Takatsuki, K., and Talal, N. (1964). *Semin. Hematol.*, **1**, 3.

Osterland, C. K., Miller, E. J., Karakawa, W., and Krause, R. M. (1966). *J. Exp. Med.*, **123**, 599.

Oxelius, V. A. (1974). *Clin. Exp. Immunol.*, **17**, 19.

Paraskewas, F., Heremans, J., and Waldenström, J. (1961). *Acta med. scand.*, **170**, 575.

Pick, A. L., and Osserman, E. F. (1968). In *Amyloidosis* (Mandema, E., Ruinen, L., Scholter, J. H., and Cohen, A. S., Excerpta), Medica, Amsterdam, p. 1.

Pincus, J. H., Jaton, J. C., Bloch, K. J., and Harboe, E. (1970a). *J. Immunol.*, **104**, 143.

Pincus, J. H., Jaton, J. C., Bloch, K. J., and Harboe, E. J. (1970b). *J. Immunol.*, **104**, 1149.

Polmar, S. H., Waldmann, T. A., Balestra, S. T., Jost, M. C., Terry, W. D. (1972). *J. Clin. Invest.*, **51**, 326.

Porter, D. D., Dixon, F. J., and Larsen, A. E. (1965). *Blood*, **25**, 736.

Potter, M., and Boyce, C. R. (1962). *Nature*, (*Lond.*), **193**, 1086.

Potter, M., Fahey, J. L., and Pilgrim, H. I. (1957). *Proc. Soc. Exp. Biol. Med.*, **94**, 372.

Pritchard, H., Riddaway, J., and Micklem, H. S. (1973). *Clin. Exp. Immunol.*, **13**, 125.

Pruzanski, W., and Watt, F. (1972). *Ann. Intern. Med.*, **77**, 813.

Pruzanski, W., Platts, E., and Ogryzlo, M. A. (1969). *Am. J. Med.*, **47**, 60.

Radl, J. (1974). In *Progress in Immunology II*, Vol. 3, (Brent, L., and Holborow, J., eds.), North Holland Publishing Co., Amsterdam, p. 390.

Radl, J., Sepers, J. M., Skvaril, F., Morell, A., Hijmans, W. (1975). *Clin. Exp. Immunol.*, **22**, 84.

Ramasamy, R., Secher, D. S., and Adetugbo, K. (1975). *Nature*, (*Lond.*), **253**, 656.

Rask-Nielsen, R., Gormsen, H., and Clausen, J. (1959). *J. Nat. Cancer Inst.*, **22**, 509.

Ratcliff, P., Soothill, J. F., and Stanworth, D. R. (1963). *Clin. Chim. Acta*, **8**, 91.
Reisen, W. (1975). *Biochemistry*, **14**, 1052.
Rivat, L., Ropartz, C., Burtin, P., and Karitzky, D. (1969). *Vox Sang.*, **17**, 5.
Roberts-Thompson, P. J., Mason, D., and MacLennan, I. C. M. (1976). *Brit. J. Haematol.*, **33**, 117.
Rockey, J. H., Hanson, L. A., and Heremans, J. F. (1964). *J. Lab. Clin. Med.*, **63**, 205.

Saha, A., Chowdhury, P., Sambury, S., Smart, K., and Rose, B. (1970). *J. Biol. Chem.*, **245**, 2730.
Salmon, S. E., McIntyre, D. R., and Ogawa, M. (1971). *Blood*, **37**, 696.
Salmon, S. E., and Smith, B. A. (1970. *J. Clin. Invest.*, **49**, 1114.
Saluk, P. H., and Clem, W. (1975). *Immunochemistry*, **12**, 29.
Schur, P. H. (1972). *Prog. Clin. Immunol.*, **1**, 71.
Schur, P. H., Borel, H., Gelfand, E. W., Alper, C. A., and Rosen, F. C. (1970). *New Engl. J. Med.*, **283**, 631.
Scrivastava and Kohn, J. (1974). In *Progress in Immunology II*, Vol. 3, (Brent, L., and Holborow, J., eds.), North Holland Publishing Co., Amsterdam, p. 390.
Secher, D. S., Cotton, R. G. H., and Milstein, C. (1973). *FEBS Letters*, **33**, 311.
Seligmann, M., Danon, F., Hurez, D. M., Haesco, E., and Preud'homme, J. L. (1968). *Science, N.Y.*, **162**, 1396.
Seligmann, M., and Feizi, T. (1974). *Progress in Immunology II*, Vol. I, (Brent, L., and Holborow, J., eds.), North Holland Publishing Co., Amsterdam, p. 285.
Seligmann, M., Mihaesco, E., and Frangione, B. (1971). *Ann. N.Y. Acad. Sci.*, **190**, 4871.
Shakib, F. (1976). Ph.D. Thesis, University of Birmingham.
Shakib, F., Stanworth, D R., Drew, R., and Catty, D. (1975). *J. Immunol. Methods*, **8**, 17.
Shakib, F., Stanworth, D. R., Hardwicke, J., and White, R. H. R. (1977). *Clin. Exp. Immunol.*, **28**, 506.
Smith, E., Kochwa, S., and Wasserman, L. R. (1965). *Am. J. Med.*, **39**, 35.
Solomon, A., Killander, J., Grey, H. M., and Kunkel, H. G. (1966). *Science, N.Y.*, **151**, 1237.
Soothill, J. F. (1962). *Proc. Roy. Soc. Med.*, **55**, 395.
Spitz, E., Gelfand, E. W., Sheffer, A. L., and Austen, K. F. (1972). *J. Allergy Clin. Immunol.*, **49**, 337.
Stanworth, D. R. (1973*a*). In *Handbook of Experimental Immunology*, (Weir, D. M., ed.) 2nd edn. Blackwell, Oxford, p. 9.1.
Stanworth, D. R. (1973*b*). In *Frontiers of Biology*, Vol. 28, North Holland Publishing Co., Amsterdam.
Stanworth, D. R. (1974). *Haematologia*, **8**, 299.
Stanworth, D. R., Humphrey, J., Bennich, H., and Johansson, S. G. O. (1967). *Lancet*, **ii**, 330.
Stanworth, D. R., Humphrey, J., Bennich, H., and Johansson, S. G. O. (1968). *Lancet*, **ii**, 17.
Stanworth, D. R., Johns, P., Williamson, N., Shadforth, M., Felix-Davies, D. D., and Thompson, R. A. (1977). *Lancet*, **i**, 1001.
Stanworth, D. R., and Stewart, G. A. (1975). In *Maternofoetal Transmission of Immunoglobulins*, (Hemmings, W. A., ed.), Cambridge University Press, Cambridge, p. 7.
Stanworth, D. R., and Turner, M. W. (1973). In *Handbook of Experimental Immunology*, (Weir, D. M., ed.), 2nd edn., Blackwell, Oxford, p. 10.1.
Stefani, D. V. Guser, A. I. and Mokeeva, R. A. (1973. *Immunochemistry*, **10**, 559.
Stone, M. J., and Fedak, J. E. (1974). *J. Immunology*, **113**, 1377.
Stone, M. J., and Metzger, H. (1967). *Cold Spring Harbor Symp. Quant. Biol.*, **32**, 83.

Stone, M. J., and Metzger, H. (1968). *J. Biol. Chem.*, **243**, 5977.

Stoop, J. W., Zegers, B. J. M., van der Heiden, C., and Ballieux, R. E. (1968). *Blood*, **32**, 774.

Taylor, B., Norman, A. P., Orgel, H. A., Stokes, C. R., Turner, M. W., and Soothill, J. F. (1973). *Lancet*, **ii**, 111.

Terry, W. D. (1968). In *Immunologic Deficiency Diseases in Man*, Vol. 4, (Bergsma, D., ed.), The National Foundation, New York, p. 357.

Terry, W. D., Fahey, J. L., and Steinberg, A. G. (1965). *J Exp. Med.*, **122**, 1087.

Van der Giessen, M., Rossouw, E., Algra Van-Veen, T., Van Loghern, E., Segers, B. J. M., and Sander, P. C. (1975). *Clin. Exp. Immunol.*, **21**, 501.

Virella, G., and Hobbs, J. R. (1971). *Clin. Exp. Immunol.*, **9**, 973.

Waldenström, J. (1944). *Acta med. scand.*, **117**, 216.

Waldenström, J. (1961). *Harvey Lecture Series*, **56**, 211.

Waldmann, T. A., and Strober, W. (1969). *Prog. Allergy*, **13**, 1.

Waldmann, R., Durm, M., and Broder, S. (1974). *Lancet*, **ii**, 609.

Waldmann, T. A., Polmar, S. H., Balestra, S. T., Jost, M. C., Bruce, R. M., and Terry, W. D. (1972). *J. Immunol.*, **109**, 304.

Webster, A. D. B., Loewi, G., Dourmashkin, R. D., Golding, D. N., Ward, D. J., and Asherson, G. L. (1976). *Brit. Med. J.*, **i**, 1314.

Young, V. H. (1969). *Proc. Roy. Soc. Med.*, **62**, 778.

Yount, W. J., Hong, R., Seligman, M., Good, R. A., and Kunkel, H. G. (1970). *J. Clin. Invest.*, **49**, 1957.